Review Copy

This is an imported title distributed by
Beekman Publishers Inc., 38 Hicks Street,
Brooklyn Heights, New York 11201

TITLE: PROFITABLE FARM MECHANIZATION

USA DISTRIBUTION DATE: FEB. 1976

LIST PRICE: $24.00

Profitable Farm Mechanization

Profitable Farm Mechanization

Third edition

Claude Culpin, OBE, MA, DipAgric(Cantab), FIAgrE

Author of *Farm Machinery*

Crosby Lockwood Staples   London

Granada Publishing Limited
First published in Great Britain 1959 as *Farm Mechanization Management*
by Crosby Lockwood & Son Ltd
Second edition as *Profitable Farm Mechanization* 1968

Third edition published 1975 by Crosby Lockwood Staples Frogmore
St Albans Herts and 3 Upper James Street London W1R 4BP

ISBN 0 258 96984 9

Filmset in Photon Times 12 pt by
Richard Clay (The Chaucer Press), Ltd, Bungay, Suffolk
and printed in Great Britain by
Fletcher & Son Ltd, Norwich

94091

Preface

The scope and aims of this book remain broadly the same as those of the last edition. The main objective is to provide a handy source of information and guidance on some of the main factors concerned with economic mechanization. The readership target includes farmers, and those concerned with the supply of farm machinery, as well as advisers, teachers and students. The book is intended to be complementary to the author's *Farm Machinery*, which gives more information about types of machines and their characteristics.

Tables likely to be most often needed independently of the text for reference purposes are again grouped in an Appendix. There is still a lack of reliable information on such matters as repair costs and useful life of equipment in relation to the amount of annual use, so the empirical estimates given in earlier editions have been retained, but have been revised or adapted as necessary in the light of further experience of their use. An effort has been made to avoid the natural tendency for size to increase with the continued expansion of the scope of mechanization; where possible old tables and references have been omitted or curtailed to allow the introduction of new material.

The advantages of sharing little-used equipment, especially that needed for specialized work on small farms, are becoming increasingly recognized by farmers and their organizations, but there is scope for more development in this area. The advantages of machinery syndicates are emphasized, and examples are given to show how simple costing methods may be applied in practice to indicate the effects of levels of annual use on cost per unit of work.

The broad principles of mechanization management remain unchanged, but there is nothing static about such factors as working rates of machines, wage rates or the number of workers available for farming. An effort has been made to bring basic information as nearly up-to-date as possible, and to present it in such a way that the effects of further changes can be estimated. Owing to the rapid changes that are taking place in many costs, notably that of fuel, the reader will inevitably find it necessary to make adjustments to some of the cost figures which are used as examples.

This edition is written for use in a difficult period, when Imperial units will remain in general use by farmers, but use of metric units will steadily increase. Where it is convenient and meaningful to do so, both Imperial and metric units are given. Where the Imperial figures are approximate, the metric equivalents are suitably rounded off. For example, 'about 15 inches' might be rounded off to 'about 40 cm' rather than to 'about 38 cm', which would be a more exact conversion. Where it is desired to make more accurate conversions, reference should be made to Appendix Table A.11 in which some commonly needed conversion ratios are given.

In a few instances, such as those tables in the Appendix which are concerned with working rates of field equipment, it is considered that performances in Imperial units are more useful, and that the extra space needed to give metric equivalents of such empirical figures could not be justified. Similarly, because the difference between the Imperial ton and metric tonne is small, and generally insignificant so far as approximate work rates are concerned, only the Imperial figures are given in the text.

Where it seems unlikely that SI units will be widely used by farmers in the near future, e.g. measurement of tractor power in kilowatts, the Imperial units are generally used.

I am indebted to many friends for assistance in obtaining up-to-date information. In particular, my thanks are due to Mr John Nix, Dr Ford Sturrock, Mr Vernon Baker, Mr Hugh

Kerr, Mr Ben Burgess, Mr Gordon Newman, Mr Harold Ruston, Mr Alan Rundle, Mr J. K. E. Robinson and Mrs M. E. Vale.

Claude Culpin

Silsoe, Bedford
1975

Contents

3. Mechanization and Management: Use of Efficiency Standards   49

Requirements for Good Management of Labour and Machinery. Choice of Equipment. Capital invested in Equipment. Standards for checking Mechanization Efficiency. Checking Labour Requirements. Checking Tractor Needs. Checking Machinery Costs. Partial Budgeting. Forward Planning of Machinery Replacements.

4. Work Study and Mechanization   69

Introduction. Scope and Limitations of Work Study in Farming. Method Study. Multiple Activity Charts. Critical Analysis in Method Study. Use of String Diagrams. Securing Labour Efficiency: Some General Principles. Repetition Hand Work. Ergonomics in Agriculture.

5. Farm Power   89

Power for Field Work. Types and Numbers of Tractors Needed. Tractor Utilization. Annual Use and Cost of Operation. Operating Speeds and Types of Transmission. Optional Fittings. Tractor Types. High-powered Wheeled Tractors. Efficient Tractor Operation. Power for Stationary Work.

6. Crop Production Mechanization   112

Cultivation Implements. 'Minimum' Cultivations. Rotary Cultivation. Manure Handling. Farmyard Manure Loading and Spreading. Slurry Disposal. Fertilizer Distribution and Drilling. Inter-row Hoeing. Down-the-row-Thinners. Potato Planting. Potato Chitting. Use of Transplanting Machines. Field Crop Spraying. Irrigation.

CHAPTER ONE

The Advance of Mechanization: National Aspects

Importance of Economic Mechanization. Trend of Developments in Some Foreign Countries. Recent Trends in Britain. Equipment Statistics and Surveys. National Expenditure on Mechanization. Productivity from a National Aspect. Capital investment. Some General Effects of Farm Size in Relation to Mechanization. Sharing the Use of Farm Machinery.

Importance of Economic Mechanization

Mechanization of agriculture is a progressive development of steadily increasing scope and importance. It began, centuries ago, with simple devices for harnessing the power of man himself; developed with the construction of implements and machines designed to make use of the greater power of domestic animals, notably horses; and continues with exploitation of the use of mechanical and electrical power for almost every farming task. Since the 1930s progress has been revolutionary. A rapid acceleration in the use of tractors and other engine-driven field machines has been followed by the development of a wide range of sophisticated equipment for carrying out essential operations better and more cheaply. The scope for future development is limited only by the necessity for mechanization to be economic. Already much of the new equipment includes automatic control devices, and these are certain to play an ever-increasing part in agricultural mechanization in the future. They open up whole new fields of development, such as automatic control of the environment for both crops and livestock.

Mechanization, allied to other scientific advances, has transformed the place of agriculture in the national economy. In the middle of the nineteenth century about a quarter of the working population of Britain were engaged in agriculture, and farming produced about one fifth of the country's wealth. Today, about $5\frac{1}{2}$ per cent of the gross national product is produced by less than 3 per cent of the gainfully employed population – a labour use lower than that of any other country. Current national trends are reflected in statistics dealing with the use of mechanical equipment on farms, and in the economic data provided by the annual reviews concerned with farm prices and guarantees.

This book deals with mechanization costs and efficiency chiefly from the viewpoint of individual farmers. It is, however, an advantage in studying this subject, involving as it does both technological and economic aspects, to know something of how mechanization is developing in Britain and the rest of the world. Because of the wide scope of the subject, only a passing reference can be made here to developments in other countries, but readers who wish to obtain the broadest possible understanding of likely developments should learn something of trends in countries such as the United States, Canada, New Zealand and Russia, where the pattern of progress is broadly similar, but differs in some important details.

Trend of Developments in Some Foreign Countries

In the United States, the Department of Agriculture has published much statistical information which indicates how productivity per farm worker has been steadily increased; and estimates have been made of the proportion of the improvement that is attributable to the use of power and machinery. In 1820, each farm worker produced enough agricultural products to provide for himself and 3 other persons. By 1945, aided by tractors and equipment such as combine harvesters, each worker was producing enough for himself and 13 others. Of an increased production per worker of 44 per cent between 1917–21 and 1945 it was estimated that half was due to mechanization and the rest to scientific advances which resulted in increased yields. The increase in productivity per worker is at present round about 4–5 per cent per annum. In 1955, 12·5 per cent of total man-power in U.S.A. was engaged in agriculture, and it has been estimated by the U.S. Department of Agriculture that the proportion needed by 1980 should be down to about 5·3 per cent.

In regions such as parts of the Middle West and of the Eastern States, where the farming is broadly comparable with that of Britain, American methods are worth study on account of the economic manner in which the farms are mechanized. Compared with similar British farms the Americans usually have fewer machines; yet they work these machines for long hours when the occasion demands, and achieve a high output per worker. Economy in mechanization is assisted by uniform easy-working soils and good-sized fields of regular shape. Moreover, the fact that there is often only one regular worker – the farmer himself – makes it easy to decide that only one set of equipment is needed. In addition, the Extension Service in some areas has succeeded in making farmers conscious of the need to study mechanization techniques with a view to securing increased output from a given amount of labour in the cheapest possible way. Mechanization methods attract the attention of many agricultural economists and farm management specialists, and there is extensive literature on various economic aspects of mechanization.

In Canada, conditions in much of the East and the extreme West are not unlike those in Britain, but the prairie farms are entirely different and represent, along with the adjacent 'Great Plains' area of the United States, one of the most extreme examples of mechanization that can be found anywhere in the world. Here, as in some of the Russian steppe lands, the simple alternation of cereal cropping and fallow leads to a very inexpensive form of mechanization, but suffers from serious disadvantages that are now well recognized.

New Zealand farmers contrive to achieve a high output per man by making the best of their pastures and climate, and generally providing each worker with as much equipment as he can handle for doing time-consuming chores such as milking. There is only one worker to about 155 acres (62 ha) of farm land. Extensive use is made of advanced techniques such as aerial top-dressing, in order to improve the production from areas that are difficult or impossible to deal with by tractor power. This work, as with aerial top dressing and spraying in the United States, is carried out by contract services. The further mechanization progresses into such specialized fields, the more impossible it becomes for family farmers to carry out the work with machinery of their own.

Several of the countries of Eastern Europe are of particular interest from the viewpoint of mechanization, on account of the efforts which have been made to employ nationally planned policies, through a system of very large state and/or collective farms. Such policies clearly permit rapid introduction of large-scale machinery where it is thought to be economic, sometimes in circumstances where the topography is well suited to the use of wide machines. Experience has shown, however, that national planning of mechanization, with manufacture of all machinery in government factories, and distribution, repair and contract work all carried out by government agencies, has many disadvantages. In most countries it has been found necessary to give the farms considerable freedom and incentives to plan their own programmes, including the selection of power and machinery. Compared with Britain, labour is abundant and cheap; and as a result, the introduction of mechanized methods often fails to produce the reduction in labour employed that would be possible if those released could be readily absorbed by other industries. This applies particularly to mechanization in livestock production.

One of the major obstacles to economic mechanization in many countries is the small size of farms. Though this is a quite serious problem in Britain, the situation in many countries of Western Europe is far worse, a high proportion of the farms being too small to provide a reasonable income for the occupiers in modern conditions. This is also one of the major problems in many other parts of the world, especially in parts of Africa and Asia.

Recent Trends in Britain

Any study of the mechanization of British agriculture must take account of the fact that most British farms are rather small. The distribution of numbers of holdings of various sizes, and types is shown in Table 1. Size of a farming business is

Table 1. Number of Full-time Holdings by Type of Farm in Various Size Groups in England and Wales, 1971*

Type of farming	Size Group in Standard man days (smd) 275–599	600– 1 199	1 200 and over	Total Number of holdings	Smd Thousands
Specialist dairying	14 712	11 638	3 832	30 182	22 752
Mainly dairying	7 158	7 064	4 507	18 749	18 341
Livestock rearing and fattening ⎱ Mostly cattle	3 645	1 696	561	5 902	3 903
Mostly sheep	1 720	1 200	403	3 323	2 418
Cattle and sheep	6 094	4 312	1 369	11 775	8 654
Predominantly poultry	875	962	1 302	3 139	6 760
Pigs and poultry	2 820	1 906	1 471	6 197	6 539
Cropping, mostly cereals	3 133	3 440	2 705	9 278	10 223
Cropping, general	4 338	3 917	5 586	13 841	21 591
Predominantly vegetables	730	452	447	1 629	2 270
Predominantly fruit	695	459	696	1 850	3 123
General horticulture	3 438	2 853	3 115	9 406	16 032
Mixed	3 863	4 233	3 624	11 720	15 091
Total holdings with 275 smd or more †	53 221	44 132	29 638	126 991	137 698

* Based on Ministry of Agriculture, Fisheries and Food Statistics.
† There were, in addition, 101 643 holdings with less than 275 smd, and these required a total of 9 538 thousand smd.

not necessarily closely related to the area of the holding. Holdings are classified, by methods described in Chapter 3, according to the number of standard man days needed annually for the enterprises undertaken. On this basis, a one-man holding is one requiring 275 standard man days (smds) annually; a one-to-two man business is one needing 275–599 smds annually; a two-to-four man business is equivalent to 600 to 1199 smds, and 1200 or more smds indicates a four-or-more man enterprise. Of about 127 000 full-time holdings in England and Wales, about 53 000 have a labour requirement of 275–599 standard man days, equivalent to one or two men, and can be regarded as family-size holdings; about 44 000 have a labour need equivalent to 3–4 full-time men, and can be considered medium-sized businesses; and there are about 30 000 holdings needing over 1200 smds, equivalent to 5 or more men, which are substantial commercial businesses. The small group of large commercial businesses produces about half the total output of the agricultural industry.

Table 2. Concentration of Enterprises*

Enterprise	Enterprise Size Group	Percentage of Total Enterprise in the Size Group by Year			
		1960	1965	1971	1972
Dairy cows	50 or more cows	21	30	51	55
Beef cows	50 or more cows	15	19	29	30
Breeding ewes	500 or more ewes	16	21	28	30
Breeding pigs	50 or more sows/gilts	16	25	51	55
Wheat	100 or more acres (40 or more ha)	26	42	56	58
Barley	100 or more acres (40 or more ha)	41	50	59	59
Main crop potatoes	50 or more acres (20 or more ha)	19	27	31	30
Laying fowls	1000 or more birds	25	46	86	88
Broilers	20 000 or more birds	42	71	86	87

* Based on Ministry of Agriculture, Fisheries and Food Statistics.

There are differences in farm structure between individual countries. Percentage figures for the different size-groups for England alone are similar to those for the United Kingdom, which has about 188 000 full-time holdings: but Wales alone has a higher-than-average proportion of small farms. The total cultivated area of the United Kingdom is about 30 million acres (12 million ha), and includes about 160 000 'significant farming units' which are smaller than one-man holdings. There is a trend towards fewer and larger holdings, but average rate of increase in size is only about 2 per cent per annum. The Ministry of Agriculture publishes detailed statistics on many aspects of the structure of British agriculture.[1]

Throughout British agriculture there is a trend towards concentration of enterprises, as indicated by the figures in Table 2.

[1] See List of References on page 283.

The pace of change in concentration of enterprises naturally tends to fall after an economic size is reached, or when the area of land available is a limiting factor. For example, the average annual percentage increase in the large size group for both barley and potatoes between 1968 and 1972 was less than 1 per cent; and it was only just over the 1 per cent per annum rate for large herds of beef cattle and sheep. With poultry, a high degree of concentration has already been achieved, and any further size increases are likely to be related to considerations other than the improvement of production efficiency.

Most small farms employ little hired labour, yet mechanization of some operations on family farms is no less desirable than it is on larger farms. But it is difficult on the basis of 'economics' alone to justify a high capital expenditure on the equipment of small farms unless a high output is achieved. Nevertheless, mechanization is proceeding apace on many small farms, as well as large. The problem of mechanization on small farms is returned to later (p. 14).

The relative importance of the main British farm products is indicated by Table 3. These figures show the high importance of livestock, with milk and milk products and fatstock contributing over half the total final sales.

Table 3. Farm Sales for the United Kingdom, 1972–3 *

Farm Sales (including subsidies)	£ million	Percentage of total
Milk and milk products	626	22
Fatstock:		
Cattle 538		
Sheep 124	992	35
Pigs 330		
Eggs	179	6
Poultry meat	165	6
Grain	320	11
Other farm crops	183	7
Horticultural products	333	12
Other	44	1
Total	2842	100

* Forecast.[2]

Equipment Statistics and Surveys

The Ministry of Agriculture's equipment statistics indicate the rapidly changing pattern of mechanization.

One of the advantages of such history as the statistics reveal lies in the understanding it permits of the way in which needs and standards can change in a short time. Equipment that seems a necessity today may be outmoded in 10 years time. As a general rule, however, changes are gradual and general trends can be foreseen by those who study the subject carefully.

National Expenditure on Mechanization

Tractor numbers increased rapidly during the 1950s, but later, as the number of workers was reduced, the main feature became an increase in average tractor size while numbers showed a tendency to fall. Similar trends are to be seen with combine harvesters where the capacity of some modern machines is many times greater than that of the biggest available only a few years ago. Root harvesting machines are steadily taking over operations formerly done by hand. In the case of sugar-beet harvesting, no hand work is necessary; but with potatoes there are still some difficult soil conditions where hand picking can compete with mechanical harvesters provided that the workers are available. However, even these difficult areas are being steadily eroded by electronic devices which automatically separate potatoes from stones or clods.

One of the main growth areas for machine numbers embraces equipment concerned in livestock production. Mechanized stock-feeding and manure-handling systems are rapidly replacing hand-work systems; and where the results justify the expenditure, as in pig and poultry rearing, fans and heaters are increasingly used to control the atmosphere in stock buildings. On modern dairy farms, collection and dispatch of the milk are normally press-button operations.

On horticultural holdings, mechanical sorters and conveyors are used to assist in separating produce into standard grades, and automatic weighers and packaging equipment are often used for the final farm operation before dispatch. Fork-lifts are increasingly used for internal handling as well as for loading and unloading road vehicles.

Production of farm machinery in Britain is an important industry. British manufacturers export tractors and other farm machinery worth about £400 million annually. Success in world markets helps manufacturers to provide the home market with first-class machines at a modest price. National expenditure on plant and machinery fluctuates according to a wide variety of factors, chief of which is the average farming profit being made in the particular year. Profits on a national scale are influenced by weather and market conditions. Expenditure on machinery is also influenced by changes in taxation regulations, and by the introduction or impending removal of government grants towards improvement of farming efficiency. Allowing for the sometimes wide fluctuations due to these external factors, current U.K. expenditure on new farm equipment is in the region of £200 million annually, of which over £50 million goes on tractors.

Total expenditure by farmers on mechanization is much higher than the £200 million spent on new machinery. National figures for expenditure normally include such items as fuel costs, contract work, and use of road vehicles for farming purposes. Table 4 shows the relationship between the various items making up the total cost of farm production. Machinery costs in this context include depreciation, repairs and spare parts, housing, licences and insurance of tractors and road vehicles, fuel, oil, electricity and all charges for contract work and haulage. The relationship of machinery and labour to other costs is affected by trends in the levels of farm rents, fertilizer subsidy, the number of livestock kept, and many other factors.

In the Eastern Counties, where arable and mixed farming predominate, labour and machinery together represented over 50 per cent of total expenses during the early 1950s, and the present level is a little over 40 per cent. Individual farm costs may depart widely from national or regional averages. With this reservation, labour and machinery together represent about 35 to 42 per cent of total expenditure. Labour and machinery costs tend to be low on mainly cereal farms and higher on farms with a wide range of crops such as fen-arable farms or farms with mixed cropping and livestock. A change such as

Table 4. Relative Importance of Items Making up Total Farming Expenses

	England and Wales Source: Annual Review White Paper		Eastern Counties Source: Cambridge University Dept. Rural Economy
	1972–3 (forecast)		1971 harvest
	£ million	%	%
Labour	449	20	22
Machinery	391	17	20
Total of labour and machinery	840	37	42
Feedingstuffs	706	31	20
Fertilizers and seeds	218	9	14
Rent (gross)	180	8	14
Other	334	15	10
Total	2278	100	100

intensification of livestock production, which involves increased use of requisites such as fertilizers and feedingstuffs, automatically reduces the percentage contribution of the 'fixed' costs.

When studying expenditure on agricultural production it is useful to consider machinery and labour together, since expenditure on machinery can usually be expected to result in some reduction in the need for labour, though not necessarily in the cost of it. The broad pattern of changes in the cost of labour and machinery in the Eastern Counties between the early 1930s and thirty years later are shown by a series of reports of economic studies carried out on the same farms throughout the period.[3] Table 5 summarizes the changes.

The War years and rapid re-equipment afterwards resulted in the high figures for 1945 to 1950. Since that time there has been a period of steadily increasing efficiency, during which the total of machinery and labour costs has remained fairly steady, while the amount spent on other items such as feedingstuffs has been rising. A continuation of such improvements may be

Table 5. Labour and Machinery as Percentage of Total Farming Costs in the Eastern Counties

Year	1931	1935	1940	1945	1950	1955	1960	1964	1971
Labour	37	32	28	36	33	27	24	24	22
Machinery	10	10	13	19	24	20	18	20	20
Total	47	42	41	55	57	47	42	44	42

expected to require further investment of capital, and this is considered later. In the meantime, it should be noted that average capital investment on Eastern Counties farms, which was only £2 per acre (£5/ha) in the thirties, had risen to about £18 per acre (£45/ha) by the early 1970s. Average annual expenditure in the Eastern Counties in 1970 was about £13 per acre (£33/ha) for machinery and £15 per acre (£38/ha) for labour.

Productivity from a National Aspect

Effects of the increase in mechanization on productivity of labour in Britain are difficult to calculate. Estimates made during the early post-war years revealed a position which caused uneasiness in official quarters, and some doubt as to whether the rapid mechanization taking place was really justified by the results. The increase in output per worker between 1939 and 1953 was estimated at about 30 per cent; but when allowance was made for various scientific advances, the increase attributable to mechanization was no more than 15 per cent. When it is recalled how great was the machinery cost needed to secure this modest increase, the doubts which were frequently expressed in the words 'over-mechanization' are understandable.

A study of changes on a restricted sample of farms gives a more useful picture than can be obtained from national statistics. Comparisons made by the Cambridge Farm Economics Branch on 20 identical farms in East Anglia showed that improved cultivations, crop varieties, fertilizer practices and other scientific advances combined to result in crop yields that were one-third higher in 1955 than they had been in 1935. The cost of labour (adjusted for changes in money values) was lower by about 13 per cent, and considering labour alone, net output per £100 labour rose by 62 per cent during the 20 years. Owing to the substantial increase in machinery costs, however, output per £100 total labour and machinery cost increased by only about 20 per cent — a finding in general agreement with the national figures for a shorter period referred to above. These figures took no account of such benefits as shorter hours of work and longer holidays, nor of such factors as the elimination of drudgery.

The true effects of mechanization are better indicated by studies which do not give undue weight to the post-War period, when there was a marked lag between rapid expansion of mechanization, and improved productivity. Sturrock[4] has compared methods and results on a 500-acre clay farm in 1963 with those in 1933, when the farm had just bought its second tractor but still had 9 working horses. In 1963 it had 8 tractors, a combine harvester, a baler, grain drier, etc., and employed 8 fewer men. Cropping and stock policy had changed little. Labour and machinery costs were estimated as follows:

	Actual 1933	1933 at 1963 prices	Actual 1963
Total labour cost (£)	1057	7086	3968
Total machinery cost (£)	333	1381	3260

Profit was £170 in 1933 and £5139 in 1963; but it was estimated that if farming in 1963 had been done exactly as in 1933 there would have been a loss of £2700. 35 per cent of the improvement in the 30 years was reckoned to be due to mechanization, and 53 per cent to scientific improvements such as new cereal varieties and greater use of fertilizers.

The average increase in labour productivity during the 1960s and early 1970s was maintained at about 6 per cent per annum. This was double the rate for the national economy as a whole. During the early 1970s the rate of annual decrease in the number of whole-time agricultural workers was about 3 per cent per annum (Table 6).

Table 6. Regular Whole-time Workers in Agriculture

Year	United Kingdom	England and Wales	
1950	717 000		575 000
1962	459 000		368 000
		Employees	Family workers
1970*	273 800	167 000	50 600
1971	266 800	162 700	50 700
1972	258 200	157 600	48 900
1973	252 100	155 500	47 900
1974†		149 600	42 100

* Numbers from 1970 onwards are not exactly comparable with those for earlier years owing to inclusion of managerial and secretarial workers, definition of 'regular whole time', etc.
† Provisional.

Capital Investment

In mainly arable farming, figures for capital investment per acre or hectare give an indication of levels of mechanization; but in making comparisons it is essential to take care to avoid this yardstick when comparing farms which are not similar in all major respects. Such figures can be usefully applied to studies of changes taking place over a period of years on identical farms. In one such comparison of 20 identical farms in East Anglia over the period 1935–55, investment in machinery rose from £2 per acre (£5/ha) in 1935 to £11 (£27/ha) in 1955. In a further study of 80 farms covering the period 1953–63 the increase was from £7–8 to £15 per acre. Average farm size increased during the 10 years from 179 to 192 acres (72 to 77 ha). On the farms below 100 acres (average size 60 acres (24 ha)) investment was considerably higher, at £20 per acre (£50/ha); but there was little difference in investment per acre between the medium sized and large farms, both being in the region of £14 per acre (£35/ha). Capital investment increased much more than annual machinery costs. The additional investment in machinery on this sample of farms averaged about £1000, and this produced a satisfactory return, calculated as follows:

Average reduction in annual cost of labour	£850
Extra annual cost of machinery	£325
Balance attributable to mechanization	£525

Average investment on the 371 farms in the Farm Management Survey in the Eastern Counties in 1971 was just over £18 per acre (£45/ha), and will have increased much further with subsequent rises in machinery prices.

It should be noted that the equipment valuations referred to above represent 'written down' figures used for assessing the position of the farm as a going concern. Actual values may be higher, and the cost of replacing the equipment by a new set would be about double. On the other hand, the equipment would probably not realize more than its valuation if it were disposed of at a machinery auction.

As capital investment in mechanization by individual farmers increases, finding the money to buy new machines becomes more difficult. Though use of hire purchase facilities is increasing, and is one way of acquiring equipment that is urgently needed, this is not an attractive way for most farmers, since hire purchase is always a rather expensive way of borrowing money. Financing of machinery supply is considered in Chapter 2.

Some General Effects of Farm Size in Relation to Mechanization

Size of the enterprise influences all aspects of mechanization, and many surveys have shown how much easier it is to achieve economic mechanization on large farms than on small. Most farm implements and machines have only a short working season, and cannot be fully utilized on small farms, where the area to be dealt with or the size of the job to be done is often very small. Most machines cannot be scaled down in size and price to suit small farms. For example, a potato digger cannot take less than one row at a time, and a pick-up baler cannot take less than a single swath or windrow. The result is that even if the smallest available equipment is chosen, the valuation of implements and machinery per unit area on small farms always tends to be higher than on large, and can be excessive unless some of the work is done by equipment not owned by the farmer. In practice, small farmers in general contrive remarkably well to make the best of their situation by such methods as the use of second-hand machinery, and by occasional resort to hiring, borrowing or contract work; and differences in investment levels between large and small farms are very much less than they would be if small farmers bought machinery regardless of economic considerations.

Surveys on the relationship between farm size and mechanization efficiency show that area cultivated per tractor increases appreciably up to about 250 acres (100 ha); but account has to be taken of the fact that large farms tend to have high-powered and new tractors, while small ones generally use smaller and older power units. Studies of the relation between farm size and such factors as profit per unit area, and net output per £100 labour cost, or per £100 labour and machinery cost, show that though the inherent disadvantages of farming small areas can to a certain extent be overcome by intensifying production of specialist crops or livestock, there is a strong case on grounds of efficient use of labour and machinery for farming units of moderate or large size. As mechanization progresses there is a tendency for advanced types of field equipment to be too expensive for economic use on small farms, but fortunately there are also developments such as the introduction of small, fully automatic electrically operated food preparing units which make it possible for small stock farms to compete with larger ones in productivity per worker.

Sharing the Use of Farm Machinery

A possible solution of small-farm mechanization difficulties which has not yet received sufficient attention from British farmers is that of sharing the cost and use of equipment. When sharing machinery is suggested to a group of farmers, the advantages are apt to be lost sight of, and much is made of the difficulties of making satisfactory agreements. One of the most frequent criticisms of equipment-sharing schemes is the statement that all farmers need the equipment at the same time, and that therefore it is impossible for all to be satisfied. In practice, however, results show that with good will, arrangements satisfactory to all participants can be made. The problem is, after all, no different in essentials from that which faces the user of a similar machine on a large farm. Some fields may need urgent attention, but others can be safely left. Provided that the scheme is basically sound it is not beyond the powers of most groups of farmers to agree on a programme. Reports on syndicates[5] show that combine harvesters, pick-up balers, manure spreaders, seed cleaners and other machines can be successfully shared. Credit is obtained on favourable terms through Syndicate Credits Ltd., a non-profit-making subsidiary set up for the purpose by the National Farmers' Union.

Arrangements are made for a local machinery dealer to supervise maintenance and to provide storage and a thorough overhaul at the end of the season. Each machine is normally the responsibility of one member of the syndicate, but the dealer checks servicing at regular intervals. A responsible operator goes with the machine wherever it works, and this operator may receive a bonus if the service engineer's report on the condition of the machine is good.

Detailed arrangements for sharing cost and use naturally vary according to need. In some cases, two or more farmers will share both cost and use equally, irrespective of the sizes of their farms or the amount of work they have to be done. In such cases, after all participants have had a substantially equal share of work done, and the work is complete on some farms, further work may be done on one or more farms on a suitable agreed basis. Where three or four farmers share a machine such as a combine harvester there is usually an agreement that each shall be allowed to harvest only an agreed amount on the first round. Reports on the working of these schemes are almost entirely favourable, and a considerable extension seems likely when the advantages are fully appreciated by farmers. Syndicates are further discussed from the viewpoint of individual farmers in Chapter 2.

CHAPTER TWO

Individual Farmers' Mechanization Problems: General Considerations

Mechanization Management in Relation to Overall Farm Management. Gross Margins. Objects of Mechanization. Calculation of Operating Costs. Contract Work. Machinery Syndicates. Income Tax and Mechanization. Investment Incentives. Financing of Machinery Supply.

Mechanization Management in Relation to Overall Farm Management

One of the main objectives of farm management, of which mechanization management forms a part, is to plan for maximum profit, preferably consistent with good husbandry, so as to ensure that conditions for profit-making are not made any less favourable in the future. There are, of course, many other objectives, including improvement of living and working conditions, and provision of leisure for the family.

In practice, it is always essential, when beginning a study of mechanization management problems on an individual farm, to ensure that they are being tackled within the right overall farm management framework. This means, briefly, that mechanization problems will be studied in relation to forward-looking decisions on such vital questions as how much of each commodity will be produced, and how the various enterprises of the farm will knit together to form an efficient farming enterprise viewed as a whole.

No attempt is made in this book to describe or discuss the many aspects of farm management which are not directly related to mechanization. It is necessary, however, to refer briefly to one of the modern management techniques, viz.

'Gross Margin' analysis, since the principles are widely employed. A fuller explanation should be sought elsewhere.[6]

Gross Margins

The objective of a gross margin analysis is to provide adequate planning data without the necessity to attempt full cost accounting, which has been found in the past to be generally unsuited to use as a tool in practical farm management, partly because most farmers cannot possibly keep sufficiently detailed records to show how the various resources should be allocated. For the purpose of the analysis, it is assumed that all farming costs may be divided into two groups, viz. those ('fixed' costs) which are not directly related to the scale of particular enterprises, and those ('variable' costs) which depend directly on the scale of particular items of production. The generally accepted division is as follows:

FIXED COSTS; include rent, regular labour and all general overhead costs, including all machinery depreciation, repairs and fuel costs.

VARIABLE COSTS; include seed, fertilizers, casual labour, contract services and sprays directly related to the crop enterprise; and livestock purchases, purchased feedingstuffs, etc. directly related to the particular livestock enterprise which is being studied.

For any enterprise:

GROSS MARGIN = GROSS OUTPUT MINUS VARIABLE COST

The technique may be simply illustrated by reference to wheat growing where the yield is 1·5 ton per acre (3·75 ton/ha); price is £50 per ton; seed costs £8 per acre (£20/ha); fertilizer £8 per acre (£20/ha); and spraying etc. £6 per acre (£15/ha).

GROSS OUTPUT = 1·5 ton @ £50 = £75 per acre (£187/ha)
VARIABLE COST   = £22 per acre (£55/ha)
GROSS MARGIN   = £53 per acre (£132/ha)

If variable costs remain the same, an increased yield of 0·2 ton per acre (0·5 ton/ha) will increase the gross margin by £10 per acre (£25/ha); and as in all crop production, gross margin is in fact a very variable margin which reflects the farmer's skill as a cultivator, and in all of the day to day problems of practical farming.

A comparison of gross margins for different enterprises facilitates decisions on the choice of enterprises. Figures for gross margins for the individual farm may also be compared with those achieved elsewhere, and can help to indicate the existence of management faults which changes in technique may help to remedy.

There is no conflict of principle between the assumption made above, that the main costs of mechanization can often be regarded as 'fixed', and the fact that most of this book is devoted to study of the various changes, some of them drastic, which can be made in these costs. When drastic changes in mechanization are directly related to a particular enterprise, the simple gross margin technique has to be abandoned in favour of a detailed assessment of the changes in the total cost of labour and machinery. All that need be added is a warning against indiscriminate use of the simple gross margin technique in such circumstances.

Little need be said here concerning the relationship of more ambitious programme planning methods to mechanization management. As a general rule, the more complex techniques, such as 'linear programming', where a computer is employed to find the optimum combination of enterprises, have hitherto been used mainly to study programmes in which it is assumed that production techniques are not substantially altered. These methods are therefore at present of little interest so far as mechanization management is concerned.

'Partial budgeting', on the other hand, is a simple form of marginal planning which is of great value in mechanization management, and examples are given later to illustrate how it may be employed for such calculations as finding the expected annual cost of owning and using new equipment; the savings in labour that may be expected; and any other effects that there may be, such as a change in the value of the product.

Objects of Mechanization

For purposes of analysis, the many aims of mechanization are discussed separately below, but in practice profitability is usually influenced by a combination of many factors, and one of the commonest mistakes is to consider only one or two points and to overlook some of the important indirect consequences of any change in mechanization policy.

The farmer's cash return is not the only element that needs consideration. Many other factors are involved, and one or more of these may be decisive, even though its effect cannot be measured on a cash basis. Among the imponderable factors are family pressure for improved working and living conditions; the desire to reduce management worries by ironing out difficult labour peaks, and such partly psychological factors as the influence of taxation. Thus, a farmer may quite justifiably decide to invest in a piece of equipment which will actually result in some financial loss if there are substantial benefits in the form of decreased drudgery, increased leisure, or contented workmen. It is therefore impossible to state at all definitely what is the optimum level of investment in equipment; for neither considerations of finance nor of the physical capacity of the machines concerned can give the answer to the farmer's individual problem.

Saving Labour

Reference has been made in Chapter 1 to the broad effects of mechanization in increasing output per unit of labour, or per unit of labour and machinery combined; and standards by which a farmer can judge his own efficiency are discussed in Chapter 3.

It should be noted at the outset that saving of labour does not necessarily result in a saving of cost. Most farms are small, and employ only a few workers. Saving a part of one man's time, or managing with one less man for a part of the year, may often have no effect on the minimum number of men needed to run the farm. Nevertheless, with good planning, productive use can often be made of labour saved; and though it is only when a man's labour can be saved throughout the year that big savings of cost ensue, smaller labour-saving benefits should not be dismissed as worthless until some thought has been given to ways of utilizing the potential value of the labour.

The main scope for direct saving of labour costs lies in operations such as potato harvesting, where casual labour is employed for picking. The average amount of casual labour needed for hand picking, in addition to the regular labour needed for driving tractors, is in the region of 50–60 labour hours per acre (125–150 labour hours/ha). By comparison, the saving from use of mechanical potato planters is small – only about 8–12 labour hours per acre (20–30 labour hours/ha). The amount of labour that can be saved by a machine such as a potato harvester depends on the efficiency of its use. Actual savings may be appreciably less than the total amount of the hand-picking labour because of the need for hand work on the harvester, and the slow operating speed; but in favourable conditions the harvester may work fast, with only a single supervisor, and it also saves the labour needed for loading after hand picking. Surveys show that efficiency of use of complex equipment often increases greatly as experience is acquired and that farmers who are accustomed to using one type of complex equipment can quickly learn to get the best performance out of others.

A study of the growing of sugar beet carried out by the Cambridge University Farm Economics Department at a time when hand thinning and hand topping were still widely practised, showed a labour-need of 180–200 man hours per acre (450–500 man hr/ha) where manure spreading, singling, topping, loading and carting were carried out without the help of specialized machines. By comparison, just over 40 man hours per acre (100 man hr/ha) were needed where the then-modern equipment was used. The general adoption of such labour-saving equipment resulted in the great reduction in numbers of workers already noted. Today, there is much less scope for further substantial saving of labour, and the main problem on individual farms is getting rid of difficult labour peaks at particular seasons. There are many instances where a farmer is glad to reduce work even if he does not save money by so doing. There are also instances where a labour-saving development may be of such importance that it pays the farmer to alter his farming system in order to take full advantage of it.

The extent to which total labour demands may be evened out by a judicious combination of crops and stock, and by reducing odd jobs to a minimum during busy seasons, is well indicated by Fig. 1. It should be noted that whereas the harvests of silage, hay and corn account for the relatively low peaks on the mixed dairy farms in Holderness, it is the root harvest in October that causes most difficulty on the arable farms of the Vale of York.

A simple record of the weekly or fortnightly labour requirements, which enables charts similar to those shown in Fig. 1 to be plotted, is clearly of considerable value to a farmer who wishes to improve his overall efficiency. It makes a good starting-point for properly planned mechanization, and is worth the small amount of trouble required to keep and extract the records.

A method of calculating the 'gang-work days' required at different seasons for various enterprises is discussed in Chapter 3. This offers one simple way of beginning a study of seasonal requirements on individual farms.

Timeliness and Improved Farming Standards

One of the most important advantages of adequate mechanization is the freedom that it gives the farmer to plan his work well ahead, and keep to a reasonable time-table with the principal farming operations. The overall improvement that has been achieved in this respect in Britain since 1939 has been substantial. Today, farmers as a whole are very much more on top of their job than hitherto. Standards of cultivation have, on the whole, greatly improved as more power has become available for the work. Deep cultivations where necessary; many cultivations in a short time; operations such as subsoiling, which were formerly largely impracticable, and new techniques such as spraying for weed control, along with allied advances, such as pest and disease control and improved manuring, have resulted in better crops. While the general picture is one of improvement in timeliness, there are many farms where there is scope for further improvement.

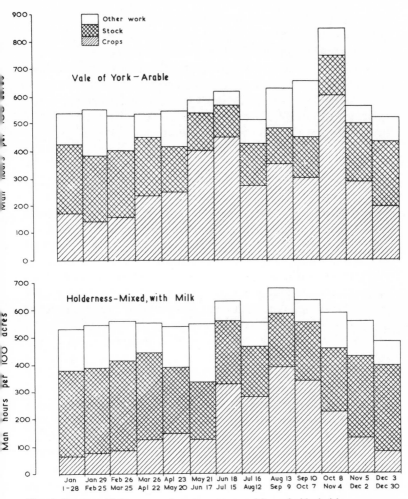

Fig. 1. Seasonal labour need on two types of farm in Yorkshire.
Source: Leeds Univ. Dept. Agric. Economics.

The important effects of timeliness on crop yields have been amply demonstrated by the results of a series of experiments on the Experimental Husbandry Farms, on the date of sowing winter wheat, briefly described in Chapter 6. It is extremely important on a corn-growing farm to be able to get winter wheat drilling done between early October and mid-November. In Chapter 3, in connection with 'gang-work days', consideration is given to the number of working days that are likely to be 'available' during a given period at different times of the year.

The achievement of timeliness may be influenced by small details of maintenance as well as by the equipment available. For instance, the immediate replacement of a broken part, either in the farm workshop or by means of a car journey to the agricultural machinery repairer, may make the difference between little or no crop and a good one.

Increased Output and Improved Quality

Increased output from a given labour force is often a sound justification for mechanization. This has been illustrated on many occasions in enterprises such as dairying. Provision of suitable equipment, such as a flail forage harvester suitable for making silage with a small gang, and a milking parlour which minimizes the time that has to be spent on milking, has enabled the number of cows handled per worker to be steadily increased. Whereas about 20 cows per worker was considered satisfactory in the thirties, twice this number is no longer satisfactory, and many farmers have 60–70 cows or more per man. On arable farms, especially where horticultural crops are grown, the quick removal of one crop will permit a valuable catch crop to be sown and harvested. Buying the machine is, of course, merely the first step: getting the best out of it requires enlightened management, and it is of the utmost importance that cropping and stocking policy and machinery management go hand in hand. Attempts to departmentalize machinery management on large farms are not always successful owing to the fact that machinery management may become an end in itself rather than a means to overall efficiency.

The possibilities of improving the quality of crops (or of milk) through adequate mechanization are so well known as to need little comment. One of the most familiar examples is the grass crop, the quality of which, whether as hay, silage or dried grass, is always dependent on speedy cutting and collection at the right stage of growth. Later, speed in haymaking often makes the difference between good and bad hay, or between moderate hay and a ruined crop. Quite apart from timeliness, the machine and technique used may have a direct influence on quality, e.g. the effect of drying techniques on the quality of malting barley and hops; and the effect of harvesting and storage methods on the quality of potatoes and apples.

Eliminating Drudgery

In a world of mechanical progress, and in a country such as Britain, farming cannot stand still and prosper. Both on the family farm and on the farm where regular paid workers are employed, machines which will reduce physical energy require-ments and render the work less distasteful to the operators are now often essential. At harvest time, men do not enjoy the heavy work involved in handling sacks of corn after bagger type combine harvesters, or handling large numbers of straw bales without the help of mechanical aids such as hydraulic loaders. It may be difficult to justify replacing a serviceable old bagger type combine by a new tanker on the basis of compar-ing increased overhead costs with the labour saved, but these are not the only factors that have to be taken into account. With bulk handling, where no heavy work is involved, the combine can be kept at work for long hours when conditions are good, and the whole job can be carried out by a small regular staff, some of whom may be incapable of doing heavy manual work. Similar considerations apply to bale handling and many other common farm jobs.

It is unnecessary here to attempt to detail all of the many commonplace mechanized operations in which elimination of drudgery is one of the important objectives. Examples include manure handling by mechanical loader and spreader, fertilizer handling in bulk rather than in bags, harvesting sugar beet by fully mechanized methods, and such regular operations con-nected with livestock as milking by machines that deliver the product directly to a bulk milk tank.

Undertaking New Jobs

Mechanization can influence cropping in many ways; and, indeed, the best results cannot be achieved unless cropping is in some measure adjusted to make the best of the equipment available. Today, equipment such as spacing drills, with selective herbicides, may make it practicable to grow vegetables where such crops were previously unattractive, while the green pea thresher makes it a reasonable proposition to grow peas for processing many miles away from the factory. Other examples include the extension of cultivation to marginal lands, an extension which often only becomes economic if the work can be efficiently mechanized.

In some instances, the purchase of machinery may be justified because it reduces dependence on the work of contractors. Thus, the marketing of a cheap low-volume spraying machine may prompt a farmer to do his own spraying for weed control in corn crops. An advantage of such a purchase will be that the work can be done at the right stage – something that the best of contractors cannot achieve for all their clients.

Reduction of Operating Cost

Since this book is so largely concerned with improvements in costs and profits by means of mechanization, it need hardly be said here that reduction of operating costs is often a prime objective of mechanization. In arable farming, there are circumstances in which the introduction of mechanized methods can clearly result in reduced total production costs. The way in which this is brought about is, of course, usually by reducing the labour required; and in instances where a number of operations are telescoped, as in beet harvesting, or where many laborious operations are completely eliminated, as in changing from hand work to a green pea harvester, the results in terms of reduced cost can be substantial.

Calculation of Operating Costs

A farmer often needs to know what it will cost him to use a particular machine. In some cases he may be considering changing from one type of machine to an improved one, or he may wish to know what the cost of using his own machine will be in comparison with the cost of contract work. As pointed out in Chapter 1, it is often convenient for neighbouring farmers to share the cost and use of one or more machines, and in such cases it is necessary for those concerned to understand the principles of costing so that equitable arrangements for co-operation can be worked out. For whatever reason costs are required, the principles are the same, and if these are understood and some basic facts and assumptions which will be considered later can be agreed, it is a simple matter to adjust the costing system to suit the particular need.

The factors involved in calculation of the operating costs of farm machinery are as follows:

(*a*) Depreciation. This charge is related to the capital cost of the machine and covers the fall in value until the machine is either worn out or obsolete, or is sold.
(*b*) Interest on capital invested in the machine.
(*c*) Miscellaneous charges for housing, insurance, etc.
(*d*) Running costs. (Fuel, oil, etc.)
(*e*) Repairs and Spare Parts.

Many farm machines are so little used each year that depreciation is not much influenced by the actual annual usage. In such a case the annual depreciation charge may be considered as a fixed sum which is decided only by the initial cost, the estimated life of the machine, and the method by which annual depreciation is calculated. Interest on capital and charges for housing and insurance are also largely independent of the amount of annual use, and it is often convenient to consider operating costs as being made up of two distinct parts, viz.

(1) Fixed Costs. (Depreciation, Interest, Housing and Insurance)
(2) Variable Costs. (Fuel, Oil, Repairs, Spares, etc.)

The division into fixed costs, which are independent of the extent of use, and variable costs, which are more or less directly proportional to the amount of use, is not an absolute one. As will be shown later, there is much to be said for considering depreciation charges as being made up of two main parts, one of which (including obsolescence) is independent of annual usage, while the other part, made up chiefly of 'wear and tear' is considerably influenced by the amount of use to which the machine is put.

This division into 'fixed' and 'variable' costs is not the same as that employed in gross margin analysis, briefly described earlier. The purposes are, however, different. In gross margin analysis the objective is usually to decide the best combination of enterprises on the assumption that mechanization is not greatly affected. The machinery costing methods now being discussed, on the other hand, together with the technique of partial budgeting discussed later, are expressly designed to enable comparisons to be made between different systems of mechanization, generally within a framework of overall farm management which has already been provisionally settled.

It is now necessary to consider these costs in detail.

Depreciation Charges and Useful Life of Equipment

Before considering the best basis for depreciation charges it is necessary to consider generally the problem of assessing the annual use and useful life of machinery. These factors are very variable in different conditions and for slightly different machines. Machinery depreciates in value from the moment that it comes on to the farm. The rate of depreciation varies according to the kind of equipment, the amount of work that it does and the care that is taken in servicing and storing it.

For some purposes, such as the calculation of expenses to set against income for income-tax purposes, it is legitimate to use certain fixed rates of depreciation which take no account of the annual use of the equipment or of any of the other factors that might influence a machine's useful life. Such methods are, however, largely unrealistic if it is desired to compare the costs of different machines or methods. For example, it is commonly assumed that the average working life of a medium-powered wheeled tractor is about eight years. In fact, however, there are many machines more than double that age still going strong, merely because they have spent most of their lives on farms where they are well cared for and little used. If the average life of a tractor must be estimated, a figure of 10 000 working hours would be a more reasonable guess than eight years; and this might represent a working life of eight years at 1250 hours per annum, or one of twelve years at about 800 hours per annum. With care in maintaining and overhauling, the tractor may continue to give good service well beyond the 10 000 hours mark. This applies particularly to tractors which have a very high annual use. For example, the life of a tractor which does 2500 hours a year and is well cared for can confidently be reckoned at 5 years. Little-used tractors, on the other hand, may well become obsolete before they have done 10 000 hours.

With most farm machines other than tractors it would be unwise to assume the likelihood of a life of over 12 years, even if the machine were used very little. Some depreciation takes place with time, regardless of usage. This type of depreciation varies considerably according to the machine's construction and the care that is taken in housing and generally maintaining it. For example, unprotected sheet metal will corrode rapidly if not kept dry, and expensive rubber tyres can be ruined by being left standing deflated with the weight of the machine on them.

Quite apart from this factor of depreciation with time, there is the factor of obsolescence, which can be important in the case of little-used equipment. Some machines are so little used that they become obsolete long before they are worn out. Obsolescence is a factor of particular importance in the case of machines which are still being rapidly improved. Such machines may be worth little or nothing as soon as they are superseded by a machine incorporating a mechanism which does the particular job much better. In all such cases, therefore, it is wise for the farmer to make a conservative estimate of the useful life of the equipment. Obsolescence of machinery can also be caused by changes in farming policy. Such changes may be necessitated by factors outside the farmer's control, or may become desirable owing to changing price levels. Clearly, any specialized little-used machine should not for costing purposes be given an estimated life longer than say 10 years unless the farmer is fairly certain that he will continue to need and to use the machine in the distant future.

The life of machinery can be determined to a certain extent by means of broadly based surveys covering machines recently scrapped or partly worn out. In order to be reliable, however, such surveys must be reasonably up-to-date, since machines are always being improved. There are, unfortunately, few survey results which can be used to assess the life of British equipment, and though many such surveys have been carried out in U.S.A., the results are conflicting. On the whole, American figures tend to show annual hours of use very low and total life high. A study by the United States Department of Agriculture showed that annual use of machinery in the late 1950s had become appreciably less than it had been for similar machines ten or fifteen years earlier. This in some cases resulted in a longer working life. For example, average tractor life was estimated at over 16 years compared with about 12 years in an earlier study. Modern American views, however, generally favour the assumption that most machines will be obsolete before they are 20 years old, and that machines fairly fully used should only be assumed to have a relatively short life. In a list giving life and repair cost issued by the American Society of Agricultural Engineers potato and sugar beet harvesters, balers, all kinds of tillage tools, front loaders and manure spreaders are given a useful life of 2500 hours, while life of combine harvesters, forage harvesters and mowers is put at 2000 hours and of sprayers and fertilizer equipment at 1200 hours. Tractor life is estimated at 12 000 hours. An authoritative text-book recommends depreciating balers, combine harvesters and forage harvesters over a 10-year period when annual use does not exceed 250 hours. The same authors recommend that less complex types of harvesting equipment, tillage and planting equipment and practically everything else should be depreciated over 15 years except where annual use is very high.

Average depreciation rates may be expressed by assessing the remaining value of the equipment at various ages as a percentage of the list price. Typical American figures for remaining value on this basis are:

At beginning of year	Wheel tractors	Combine harvesters	Forage harvesters and balers	All others
2	63%	59%	52%	55%
10	36%	25%	20%	23%

There are several factors to be considered when estimating annual depreciation of a particular machine. As a general rule, good management requires that the rate of writing off capital invested cannot be allowed to depend only on the working life of the machine. Yet the assumption that the 'depreciation life' is a fixed small number of years, with no allowance for annual usage, is also clearly so far from the truth that it should only be made in exceptional circumstances. There is therefore much to be said for a compromise, with depreciation life related to annual use, but with the proviso that no machine should be depreciated over a period of more than 12 years, even if annual use is very small, and physical life is likely to be considerably longer. This is the basis used for the figures in Table A.2(i),* in which machines are grouped into a number of different categories, and figures are given for a suitable depreciation life at various levels of annual use.

Published facts and estimates have been made use of in a few cases where information is available and thought to be applicable; but most of the figures are based mainly on general experience. It should therefore be emphasized that in circumstances where more reliable figures for depreciation life are available, these should be preferred to those given in Table A.2(i).

The figures in Table A.2(i) are the result of an attempt to take account of the particular characteristics of the different groups of machines in respect of general durability, normal wear and tear and obsolescence. Implements such as mounted ploughs, if properly maintained and not mis-used, literally never wear out; and even where use is heavy they do not have a short life, save in exceptional circumstances such as if the implement is over-strained. But their usefulness on a particular farm may be of much more limited duration, due to an increase in the power of the tractors used, or a desire to change to a newer type.

In the case of equipment such as fertilizer distributors and combine drills, the life may sometimes be short even at low annual usage, on account of rapid deterioration caused by neglect of cleaning. Though a restricted survey shows average life higher than is suggested here, more extensive evidence is

---

* All tables with prefix A are shown in the Appendix.

needed before the estimates can be set appreciably higher. Accidents or careless handling can wreck some machines in a very short time, and such factors are not allowed for in Table A.2(i). The low estimates of life given for potato harvesters are mainly influenced by the obsolescence factor.

No attempt has been made to make fine divisions because of the big variations in original quality of machines and in standards of maintenance on different farms. Thus, though the simplest mounted ploughs, which are completely devoid of mechanisms, clearly have a longer potential life than say trailed cultivators, no attempt has been made to differentiate between the two types of equipment. Readers are therefore advised to use the figures merely as a general guide, in the absence of more reliable information.

A method of estimating depreciation introduced by V. Baker incorporates the simple principle of straight-line depreciation, yet makes allowance for the age at which the machine is renewed, and so provides a rapid estimate of depreciation at any stage of a machine's life. For this purpose machines are divided into three broad classes, with those having a very high depreciation rate at one end of the scale, those with a very low rate at the other, and machines such as tractors, combines and pick-up balers intermediate. Figures for the average annual fall in value of equipment on this basis are given in Table A.2(ii). This table can help in considering questions concerning machinery replacement on individual farms.

Relationship of Depreciation Rates to Return on Capital and Re-Investment

It can be argued that capital should not be invested in equipment unless there is likely to be additional profit from the investment equivalent to the profit that can be obtained from investment of tenant's capital in farming as a whole. This reasonable concept sometimes leads to less valid statements to the effect that money should not be invested in machinery unless it gives, for example, 20 or 25 per cent per annum return on the capital; or alternatively, unless the equipment can be written off in 5 or 4 years. Sometimes even a 3-year write-off period is specified. In support of such contentions it can be shown that for selected enterprises at particular times, money invested in the essentials of crop or stock production can and often does produce additional profits corresponding to these figures. Nevertheless, it is certain that if such principles were applied to the mechanization of agriculture generally, progress would be slow, and such policies would not be tenable in the long term. The figures for depreciation rates given in Table A.2(i) are believed to be realistic for general use; but there are circumstances where it is desirable to 'write-off' equipment at much faster rates. This applies particularly to situations where tenant's capital is very limited or quite inadequate; where there is nevertheless considerable scope for expansion of output; and where increased production can be achieved cheaply, thereby securing a high return on additional money spent on such essentials as livestock, fertilizer, etc. In these circumstances there is every reason to avoid spending money on machinery or anything else that is likely to limit the size of the productive enterprise. In these conditions it is reasonable to insist that normal depreciation rates be abandoned, and that money spent on equipment be written off very quickly. Everything possible must be done to secure maximum output from a limited capital investment. There are likely to be advantages in contract work, hiring of equipment, hire-purchase, machinery syndicates or buying second-hand machinery with a view to cutting capital investment wherever possible. In such circumstances, when comparing other methods with machinery ownership, it would be sensible to assume that owned machinery should be written off much faster than indicated by the figures in Table A.2(i).

Calculation of Depreciation Charges

When the useful life of a machine has been estimated, it is necessary to decide between two main methods of computing the depreciation charge, viz. the 'inventory' method and the 'straight line' method. By the inventory method, the charge is calculated as a fixed percentage of the diminishing value. The value of the machine is written off quickly in the early stages, when repair costs may be expected to be low. It may often be a convenient method for calculating farm profits, but does not usually provide the best basis for comparing machinery operating costs. For the latter purpose, the simple 'straight line' method is usually both easier and more satisfactory. By this method the annual or hourly depreciation charge is obtained merely by dividing the capital cost of the equipment by its estimated service life. The hourly depreciation charge remains constant throughout the life of the machine, and though this does not correspond strictly to the facts, there is not a large error if repair costs are similarly evened out. For example, using Table A.2(i), the hourly depreciation charge for a £2000 tractor used 1000 hours yearly (life 10 years) is

$$\frac{£2000}{10\ 000} = £0\cdot20 \text{ per hour.}$$

Similarly, the life of a combine harvester which does 100 hours a year is put at 12 years, so if the machine costs £4800, annual depreciation charge is £400 and hourly charge £4. Simple figures such as these help to emphasize the usefulness of costings. If the combine can usefully be employed for 200 hours a year by working for a neighbour, the annual depreciation charge (life 9 years at 200 hours) is only increased to £533 and the hourly charge falls to £2·67 per hour, equivalent, at a reasonable output of $2\frac{2}{3}$ acres per hour to £1 per acre.

When detailed economic studies are made of particular types of highly developed machines there are valid arguments in favour of choosing more elaborate methods of assessing the incidence of depreciation. With machines such as tractors and combine harvesters there is a very heavy drop in re-sale value of the equipment during the early years, regardless of the level of use; and if depreciation is assessed at a constant high percentage of the diminishing value, the result accords well with actual re-sale values and is much nearer to the facts than figures obtained by the simpler straight-line method, with variations according to the amount of annual use. It is therefore considered that while the straight-line method is adequate for purposes where long-term effects of using machines are concerned, the inventory method should be preferred in circumstances where it is important to obtain a realistic assessment of the value of a few important items of equipment at various ages, e.g. when considering at what stage replacement is most economic for the individual farm.

Interest on Capital

When calculations are being made concerning overall management questions there may be good reasons for not charging interest on capital invested in farming, or alternatively for charging it at the full rate charged for interest on a bank overdraft. In support of not making a charge, it may be contended that the investment is made with a view to making a farming profit, and that the profit made is a measure of the interest which the money earns. In support of the alternative of a high charge, it can be argued that such rates must be paid on borrowed capital. Alternatively, it may be contended that the average return on tenant's capital may be of the order of 15–20 per cent, so similar charges should be levied on any capital which is invested in machinery.

In this book, which is concerned essentially with that part of management which is directly related to mechanization, it is assumed that capital for investment in machinery normally comes mainly from farming profits, and that when the cost of doing farm work by various methods is being compared it is necessary to include a charge for interest on the money which is invested in the equipment. Clearly, the farmer who has money to invest could obtain a return on it if he invested it outside farming. He could also possibly use the money to produce a higher return from farming by investing it in say extra fertilizer. Because the capital cost of the equipment is written off regularly in the form of depreciation charges, the amount of capital invested falls from that of the purchase price in the first year, to nothing when the machine is completely written off. Using the 'straight-line' depreciation method, the average investment is clearly half the capital cost, and the generally accepted method of charging interest is in fact on half the capital cost over the whole life of the machine. The interest rate can be adjusted according to whether it is necessary to consider the matter as an investment for money that is available, or whether the money needed has to be borrowed. In this book a rate of 8 per cent per annum is assumed in all examples.

Repair and Servicing Costs

Maintenance costs include some small 'fixed' charges for housing and insurance, and the charges for repairs, spare parts and servicing, which vary with the amount of use of the machine and tend to increase steadily as it grows older. As with depreciation and interest charges, it is best in calculating operating costs to even out maintenance charges over the whole life of the machine where this is practicable. Accurate estimates of repair costs are not easily obtained.

A study on 81 farms in the South-East of England[7] showed wide differences between individual farms in the incidence of repair costs, and indicated that a very large sample would be needed to arrive at positive conclusions concerning the factors most closely linked to such costs. With several machines, accumulated work was important, and there is every reason to suppose that this finding is reliable.

As in the case of estimates of a machine's useful life, American figures, though abundant, cannot be used. It is, however, of interest to note that several American agricultural economists quote annual repair costs as a percentage of the capital cost of the equipment; and for general reference purposes this method has advantages. On this basis, one authoritative source suggests the figures given in Table 7.

There are two main disadvantages in using such figures. One is the fact that annual repair cost should obviously bear some relationship to annual use. The other is that where two machines for doing the same job are compared, there may be a

Table 7. Annual Repairs, Maintenance and Lubrication (not including engine oil) as % of First Cost of American Equipment

	%		%
Mouldboard plough	7·4	Mower	4·2
Cultivator	3·8	Side delivery rake	2·5
Disc harrow	3·5	Pick-up baler (engine drive)	3·8
Spike-tooth harrow	1·1	p.t.o. combine harvester	4·0
Ridger	5·5	Engine drive combine harvester	3·5
Grain drill	2·2	Self-propelled combine harvester	3·4
Manure spreader	2·0	Forage harvester	4·5

choice between a cheap, badly constructed machine and an expensive but well constructed one. In such a case repairs should clearly not be in proportion to capital cost, but rather the reverse. This difficulty cannot easily be avoided, but the relationship between annual use and cost can be allowed for, and Table A.3 contains the author's estimates of suitable charges for different machines and periods of use in average British conditions. The method of using the figures in Table A.3 may be illustrated by an example. Suppose that a combine drill costs £600 and is used 100 hours per year. Annual cost for repairs is

$$\frac{5 \cdot 5}{100} \times £600 = £33 \text{ per year.}$$

Similarly, the annual repair and renewal cost for a plough costing £200 that is used 150 hours a year in normal soil conditions is

$$\frac{11}{100} \times £200 = £22 \text{ per year.}$$

As is well known, the wear on cultivation implements depends to a great extent on soil conditions, and some adjustment must be made where necessary for extreme variations from normal. In the case of machines such as combine harvesters, there are considerable differences between the self-propelled types, which incorporate a considerable amount of hard-wearing automotive mechanisms, and the lighter, p.t.o.-driven machines which have a higher proportion of their mechanisms subject to rapid wear. Allowance for this is made in Table A.3, but generally speaking, no similar allowances have been made for other major differences between machines – e.g. trailed or mounted; engine or p.t.o.-driven, etc., and the reader is advised to make his own adjustments where they appear to be necessary. The figures given in Table A.3 are based partly on and conform broadly to published data but have been varied where the average of a limited range of figures appears to the author likely to be misleading. The figures in Table A.3, like those in Table A.2(i), can serve as a guide where no better figures are available; but where records kept over a number of years are available and give different figures for the particular circumstances, these more precise figures should be preferred. Repair costs on little-used machines are likely to be higher per hour than those for machines that are fairly fully used. Table A.3 makes some allowance for this, but does not adequately cover the case of equipment used for under 25 hours a year. Such equipment requires special consideration according to the circumstances. In using the figures, due allowance should also be made for the 'quality' factor already mentioned. Where it is known that a high capital cost is due to the inclusion of high quality bearings, specially hardened wearing parts, etc., the repair cost percentage should be appropriately reduced. Reductions should also be made where soil-working implements are used in non-abrasive soil such as black fen.

Tractor and Labour Costs

The only other factors which need some general consideration before studying an example are the cost of tractor power and of labour. Except where otherwise specified, the labour cost for normal work carried out on the farm is reckoned throughout at 80p per hour, and all that need be added here is that the number of hours worked must normally include wasted time and time taken in travelling to and from the job. The reader should adjust where necessary. So far as tractor work is concerned it is usually quite unnecessary to work out tractor costs every time use of a tractor-operated machine is studied. Certain basic hourly costs for operating various types of tractors can be assumed, and the appropriate figures are discussed in Chapter 5. Thus, for an ordinary wheeled tractor of 45 h.p. a suitable hourly figure is 60p per hour or £1·40 per hour with its driver. These figures may be added to the hourly operating cost of the implement and machine after the latter has been calculated.

There is now such a wide range in types and sizes of tractor that it is necessary to specify the power and type when making any economic calculations. Typical costs of operating different tractor types and sizes are given in Table 12.

Total Operating Costs

As an example of the application of the costing methods outlined, the cost, per ton of manure spread, of operating a small p.t.o.-drive farmyard manure spreader costing £500 may be considered, first if used to spread 100 tons a year, and second if used to spread 500 tons a year. The overall rate of spreading, including the time taken in loading and transport, may be assumed to be approximately 8 tons per hour (see Chapter 6).

Referring to Table A.2(i), the life of the spreader in both cases is estimated to be 10 years. Table A.3 indicates annual repair charges of about 1 per cent and 4 per cent of the capital cost respectively. The annual and hourly costs are therefore as shown in Table 8.

Table 8. Cost of Operating Manure Spreader

Amount spread annually	100 tons	500 tons
	£	£
Fixed costs per annum		
Depreciation	50	50
Interest on capital	20	20
Housing etc. say	5	5
Variable machine costs		
Repairs and servicing	5	20
Total annual machine cost	80	95
Machine cost per ton	0·80	0·19
Tractor cost per ton	0·08	0·08
Labour cost per ton	0·10	0·10
Total cost per ton	0·98	0·37

The total operating cost of the machine that is little used is found to be high in comparison with that of the machine used for spreading 500 tons. On most farms, usage of such specialized machines tends to be low, and with equipment of this kind, which depreciates to a considerable extent even when not in use, there is clearly a strong case for schemes whereby use and cost of the equipment are shared between neighbours. (See Chapter 6.)

Operating costs can be expressed in a variety of ways, but as a general rule the basic figures are the annual and hourly costs, and from these can be calculated the cost per acre (or ha) or per ton, provided working rates are known.

In the example given on operation of a manure spreader, only a part of the complete operation was costed, no account being taken of the cost of the loader. Hydraulic loaders are placed in Group 3 of Table A.2(i) and in Group 5 of Table A.3, and it is a simple matter in the case of a one-man system to add the annual machine cost for the loader. Where a team job is carried out it is generally more useful to cost the complete operation, including all of the men and tractors, as well as the equipment.

Work for Neighbours

The extent to which figures calculated as above can be used for estimating the cost of work done outside the farm may now be considered. It should first be pointed out that figures calculated in this way cannot be used, without considerable adjustments, as a guide to fair contract prices. A contractor has to face serious overhead costs for slack time, arranging of contracts, etc., and has to meet heavy additional costs on account of higher basic wage rates and overtime in busy periods. His overheads for premises may be heavy. This consideration of work outside the farm is not, therefore, assumed to apply to regular contract work, but only to farmer-to-farmer hiring on the basis of mutual aid. In such a case, the main additional items beyond those already allowed for are (1) management and supervision of the equipment (2) transport of equipment and operators to and from the place of work, and (3) profit to the owner of the equipment.

Supervision of a complex machine such as a pick-up baler may involve the owner in a considerable amount of time and trouble, and should clearly be allowed for, but the cost is one which cannot be fixed according to any general rule. All transport cost should obviously be borne by the 'customer'. After costs have been calculated in the manner outlined above, the figure arrived at would represent the bare cost to the owner of doing the job. Nothing has been allowed for contingencies, and even if all goes well the owner will only just be recompensed for his trouble. The only advantage he obtains is that in the case of little-used equipment, by extending use to cover his neighbour's needs, his own depreciation charges are somewhat reduced. Clearly, therefore, if the neighbour does not provide him with some similar service, it would be reasonable for the owner to add a percentage to the total cost to ensure that he makes a small profit for his trouble. If 25 per cent is added to the calculated cost the total figure is still likely to be one that is very reasonable compared with the necessarily higher costs of regular contract services.

Contract Work

There must be few farmers today who do not employ contractors for one purpose or another; and for many jobs, such as applying toxic sprays, lime spreading or mole draining, this is often the most satisfactory and economic way of getting the work done. The further science advances and mechanization progresses the more impossible it becomes for the average farmer to have all his own equipment for every job. Operations which it is often well to have done by contract include deep ploughing, subsoiling and mole draining, where there is insufficient work to justify keeping a tracklayer; specialized spraying for pest and disease control; lime spreading where the dressings needed are heavy, and 4-wheel-drive spreader trucks can run direct from quarry to field and apply the lime at a low inclusive cost; sugar-beet harvesting, combine harvesting and pick-up baling; and generally other specialized work where the acreage is small and the equipment to do the job well is expensive. Every farmer who has had experience of contract work will know the importance of being able to rely on getting work done at the right time. Achievement of 'timeliness' depends on a close and continuing working relationship between farmer and contractor. Calling in a contractor as a last-minute emergency action can seldom be satisfactory. Contractors can only give good service on an economic basis if work is planned well ahead so that the specialized equipment is both reasonably fully used and ready on time for all customers. This requires forward booking of the service within specified periods, followed by close liaison between farmer and contractor during the working season.

Table A.10 gives a list of typical contract charges.

Machinery Syndicates

Reference has already been made in Chapter 1 to some national aspects of the development of machinery syndicates. From an individual farmer's viewpoint, participation in one or several syndicates can help to achieve several important objectives. These include:

Limitation of Capital Investment in Mechanization

Approved syndicates can borrow money on favourable terms (as little as $\frac{1}{2}$ per cent above the bank rate), partly because the syndicate members agree to 'joint and several' legal liability, and because of the precise arrangements which are made for repayment of capital within a relatively short time, e.g. 4 years. The individual member's capital contribution in any one year is thus usually limited to a quarter of his share of the cost of the machine. For example, if he arranges to have a one-fifth share of a light tractor-mounted hydraulic ditcher which costs £800, his capital contribution per year is only one-quarter of one-fifth of £800, i.e. £40. His annual payment is, of course, a little more because of interest charges and various overheads; but these charges are usually quite small, and the saving in capital investment is very worth while.

Equipment is Always Reasonably Up-to-date

At the end of four years service, the equipment referred to, which may have been the best of its type when purchased, will probably have been superseded by an improved model. The second-hand value of the machine owned by the syndicate will probably be as high as it would have been if it had done only about one-fifth as much work for the individual farmer. The individual could hardly justify replacing a machine which would have done so little work; but the syndicate-owned machine will by this time have been written off, and there will be no hesitation in trading it in and buying a new machine which is not just a replacement but is likely to be considerably better. The owner of a one-fifth share will by now have spent £160 of capital on the old machine, and his share of the trade-in allowance will probably at least cover his first capital payment on the new one.

Efficient Maintenance is Easily Arranged

Experience has shown that maintenance of equipment in a well-run syndicate is normally better than where ownership is individual. Members generally take some trouble to ensure that the machine is in good working order when it is handed over. Moreover, the arrangements usually made with the dealer for regular inspection and an annual overhaul help to ensure better-than-average servicing.

Machine Capacity is Adequately Utilized

One of the most important rules for successful operation of a machinery syndicate is avoidance of over-estimating machine capacity. While it is clearly desirable to secure reasonably full and therefore economic use, nothing can be worse than arranging an over-ambitious programme which cannot be fulfilled in good time. For example, when planning syndicate use of a combine harvester, it is advisable to budget for rather less than full use, in order to ensure that the stated requirements of all members can be met without incurring serious field losses. In the case of a machine such as a combine harvester, it is necessary when planning syndicate use to ignore the fact that it can do much more work in a good season, and to budget for only about 200 combining hours per season. This means that the syndicate area for a moderately priced machine capable of averaging 2 acres (0·8 ha) per hour should not exceed about 400 acres (160 ha). The machine may, of course, do considerably more than this in a good season if additional contract work is available.

Income Tax and Farm Mechanization

Some understanding of Income Tax procedure is necessary to a study of the economics of farm mechanization, so though this book cannot deal with the subject comprehensively, the general principles governing the application of wear and tear allowances are briefly outlined.

Capital expenditure on farm plant and machinery cannot normally be deducted in full as an expense in computing profits. It must be regarded as an investment on which wear and tear allowances may be claimed according to official scales which are varied from time to time according to government policy on investment and taxation.

In the past, 'standard' rates of wear and tear allowances were designed to be roughly in proportion to the rate at which the particular type of equipment could be expected to depreciate. These rates ranged from 15 to just over 30 per cent per annum of the written-down value of the equipment. In addition to these standard rates there were often 'initial' allowances designed to encourage investment by increasing the proportion of cost that could be written off in the year of purchase. There were also occasionally 'investment' allowances which were additional to the total of wear and tear allowances and so gave a more positive encouragement to investment on the part of farmers liable to income tax.

Under regulations introduced in 1973, farmers are allowed to write off the whole cost (100 per cent) of new machinery of any type in the year of purchase. This represents a substantial incentive to investment in years when farming profit is high. The regulations allow the farmer to claim less than 100 per cent in the year of purchase if he so elects; and because of the effects on total income tax paid over a period of years it can be advantageous in some circumstances to spread the relief so that a proportion of it can be taken in a high-income year. Where the maximum permitted allowance is not claimed in the year of purchase, the maximum relief permitted in subsequent years is restricted; it can be 25 per cent or less at the farmer's choice. The allowances for all equipment on which the 100 per cent allowance is not claimed are pooled and sales of equipment are deducted from the pool. If the total of allowances exceeds the qualifying expenditure (total of purchases less total sales), balancing charges are payable. Thus, if a farmer buys a machine for £500 and sells it for £100 after receiving wear and tear allowances totalling £350, a balancing allowance amounting to relief from tax on £50 can be claimed. If, however, he sold the machine at this stage for £200 he would be subject to a balancing charge of tax on an additional £50.

Clearly, it pays a farmer who is making a satisfactory profit to write off his machinery as rapidly as he is allowed to, and so claim the maximum relief from taxation, while building up a good set of equipment. By this means the greater part of the expenditure on equipment can be written off near to the time when it was deemed wise to invest in it.

The tax rate paid by an individual has an appreciable influence on policy concerning machinery replacements. The higher the tax rate, the less is the net cost to the farmer of buying machinery. For example, where no tax is paid the full cost of any capital expenditure is borne by the farmer. At the standard rate of taxation, less earned income relief, the net cost to the farmer works out at around 70 per cent of the amount spent; while at higher taxation rates, a smaller proportion of the cost, e.g. around 50 per cent, is eventually borne by the farmer.

The whole of the expenses on repairs and maintenance of equipment can be deducted before calculating profit, and some farmers who are skilled at running a workshop prefer to avoid spending much capital on equipment, and use mainly older machines. Such a policy makes it necessary to be ready to repair machines at short notice, and does not appeal to most farmers who are in a position to provide the necessary capital to keep their equipment more up-to-date. What an individual farmer decides to do must obviously depend not only on his skill as a repairer or repair organizer, but also on the whole capital situation and his personal inclinations.

As a general rule, it pays to tackle the most expensive repair projects in high-income years so far as is practicable.

Investment Incentives

Special provisions are made from time to time to encourage investment in agricultural machinery. To qualify for an investment grant, both the holding and the equipment have to fulfil a variety of conditions, which cannot be specified here but are set out in a free Ministry of Agriculture booklet.[8]

Financing of Machinery Supply

One of the common results of the intensification of farming is a shortage of capital which makes it necessary to consider methods of financing machinery supplies other than the simple one of paying the full capital cost at the time of purchase. Even if there is no long-term credit difficulty there may be short-term cash flow problems which make it necessary to study credit facilities.

One attractive method has already been mentioned, the Machinery Syndicates scheme, whereby the members together pay an initial 20 per cent deposit, and find the balance over 4 years. This scheme offers a valuable credit facility, quite apart from advantages of sharing cost and use.

Where individual ownership is sought, it may be possible to obtain credit by bank overdraft, bank loan account, hire-purchase or loans against securities such as a life-insurance policy. These methods need to be compared with others which do not involve machine ownership, including short-term hire, contract hire and leasing.

A bank overdraft on the current account, when available, usually costs $1\frac{1}{2}$–3 per cent above the current bank base rate. It is usually the most attractive method, with the exception of machinery syndicates, especially as the interest is charged on daily balances outstanding so that an advantage is gained from any receipts which can be paid into the account.

With a special bank loan account, the principal has to be repaid at regular intervals, as with a mortgage, and a fixed interest rate is charged on the outstanding balances.

With hire-purchase, the purchaser normally agrees to repay the capital sum and interest over a period of 2–3 years. Most finance houses require a substantial initial payment, usually about 20–25 per cent. Interest is usually charged on the whole sum borrowed over the whole period; and this means in effect that interest is paid on the borrowed capital at almost double the normal rate. A further possible disadvantage of hire-purchase is that there is little opportunity for buying at less than the 'recommended' retail price.

Some finance houses and insurance companies will advance money for machine purchase on the security of a life-insurance policy or title-deeds of the farm. Such credit may be at competitive rates, especially in comparison with hire-purchase. Insurance brokers are a normal source of information on such facilities.

With hire-purchase and the other purchase methods so far discussed the purchaser is usually assumed to have title to the machine, and can claim depreciation allowances for income tax, and also relief on the interest element of the instalments.

Hiring and Leasing

Short-term hiring is available from a few specialist firms and some machinery dealers for equipment such as tractors. A £2000 tractor is likely to cost about £40 per week, or possibly down to about £30 for longer periods.

The term contract hire is usually applied to long-term hiring, e.g. for a period of either 2 or 3 years. The equipment is hired out at a fixed monthly charge which includes maintenance and taxes, but need not include insurance. The supplier is required to maintain the machine in tyres and to carry out any necessary repairs; but the time at which expendables such as tyres are replaced is partly at the supplier's discretion, and is a point which should not be overlooked. The supplier will normally provide a replacement if extensive repairs become necessary. A typical *annual* contract hire charge for a 2-year contract period is about 35 per cent of the cost of buying the tractor new, so the system can only be attractive to the hirer if the tractor is fully used. The actual contract hire charge will in fact vary according to the likely amount and type of use, and the standard of care that is likely to be exercised by the hirer. The system has the advantage that all payments are allowable in full for income tax purposes, and that the farmer enjoys the benefit of a reliable maintenance service. The method is particularly worth considering where a high-powered tractor can be fully used for cultivations and also for work such as forage harvesting.

Leasing

Leasing differs from contract hire mainly in that the farmer is responsible for repairs and maintenance, and that a replacement machine is not normally supplied in the event of a breakdown. The terms of leasing agreements vary widely. A typical leasing period is 2 or 3 years. There is usually a substantial initial payment equal to about 4 months rent, and this may be offset by a rent-free period immediately following or at the end of the lease.

Main advantages of leasing lie in the options provided for at the end of the leasing period. These may include continuing to lease the same machine at a nominal charge, or leasing a new machine with the credit of a cash sum closely related to the value of the old machine. At the end of a 3-year lease the remaining value may be entirely the hirer's. These options usually represent substantial sums, and appeal to farmers who take good care of equipment.

Leasing charges vary with the general monetary situation. Typical charges for a 3-year lease on a £2000 tractor are 36 monthly payments of £68, which is equivalent to an annual total of 40·8 per cent of the capital cost. Comparable figures for a 2-year lease are 24 instalments of £96, and the annual cost represents 57·6 per cent of the initial value.

Dealers who run leasing schemes tend to favour sale of the machine at the end of the lease, and will often offer an attractive bargain in order to secure further business. Having recouped both capital and interest charges the dealer is happy to allow the farmer a good trade-in value if this will lead to leasing him the new machine. The farmer may by that time be happy to reduce the new leasing charge by setting the trade-in allowance against the cost of the new machine. While a lease is in operation, all of the leasing charges are normally allowable as expenses to be deducted in calculating taxable income. Leasing is therefore an attractive method of securing the use of new equipment, and it has quickly become a regular procedure for many farmers.

CHAPTER THREE

Mechanization and Management: Use of Efficiency Standards

Requirements for Good Management of Labour and Machinery. Choice of Equipment. Capital invested in Equipment. Standards for checking Mechanization Efficiency. Checking Labour Requirements. Checking Tractor Needs. Checking Machinery Costs. Partial Budgeting. Forward Planning of Machinery Replacements.

Requirements for Good Management of Labour and Machinery

The scope for improving farm profits through higher efficiency in the use of labour and machinery has been indicated in Chapter 1. This chapter is devoted to a general study of mechanization efficiency as it affects individual farmers.

Economic efficiency in farming is achieved as a result of skill and efficiency in carrying out many husbandry operations, and by applying skilled management to the whole business of farming. It is not possible in this book to review all of the many factors involved, and readers who seek information on the wider aspects of management are advised to consult a Ministry of Agriculture Handbook.[6] So far as mechanization is concerned, good management involves (1) a choice of enterprises that blend well together and are right for all the local conditions; (2) choice of a set of equipment suited to the needs of the farm; (3) the employment of sound operating techniques for using both individual machines and matched sets of equipment; (4) attention to detailed setting and adjustment to suit soil, crop and weather conditions, and (5) a well-planned system of maintenance and overhaul to ensure that the equipment is always in good working order. It need hardly be said that the last three of these items depend for their execution on a sound labour management policy. No more than a passing comment can be made on some of these aspects, but choice of equipment and operating techniques are studied in some detail.

Choice of Equipment

Surveys of labour and machinery usage on farms have shown how much tractor power and how much labour is on average employed on various types of farms for growing the main crops and tending the various forms of stock. Standards employed vary a little according to the Region and the many types of farming carried out there; so where regional rather than national standards are available and appropriate, they should be used. Such figures provide an invaluable check on mechanization efficiency, and are discussed on p. 57. In the absence of practical experience, such calculations can be used to indicate the levels at which a particular farm might be equipped and staffed for the execution of particular overall management plans. In the present state of knowledge, however, it seems advisable to employ practical experience to the full in choosing equipment, and to use the efficiency standards for checking results. The reason for this view is that the standards available are limited in scope and are rapidly made inaccurate by successive improvements in both power units and machinery. There can, however, be no disputing the need for farmers to know the approximate working capacities of the common power units and machines in their conditions, and it is an advantage if these are known both in terms of area per hour or per day, and in potential output per season. The chief difficulty is that the performance of a machine is very variable according to climatic and soil conditions, size and topography of the fields, and size and exact type of both the equipment and the power unit used to operate it. Thus, in the driest parts of England, a combine harvester of the latest design can be expected to harvest 30–40 acres of corn per foot (37–49 ha/m) of cutter-bar, while in wetter regions of the West the same machine may have difficulty in averaging 20 acres per foot (24 ha/m) annually. Even so, the problems would be relatively simple if all combine harvesters had a similar ratio between output and cutter bar width; but in practice the outputs of machines of similar cutting width may differ widely. This is a particularly difficult machine to assess, but even with ploughs, a particular implement may do 200 acres (80 ha) a year on a light-land farm where ploughing can proceed on most days of autumn, winter and spring, whereas on a heavy land farm the same implement may fail to handle 100 acres (40 ha) in the limited amount of time when conditions are suitable. With machines such as forage harvesters or pick-up balers, output is greatly influenced by crop conditions. As a general rule, outputs are appreciably higher in dry conditions than when the weather is unsettled. The amount of effective work that an implement can do clearly depends also on good planning and management. Thus, one tractor hoe can easily handle over 100 acres (40 ha) of root crops which are nicely spaced out in germination time and growth so that all fields do not need attention in the same week. If, however, 60 acres (24 ha) of sugar beet all needs hoeing at the same time and unsettled weather intervenes, one hoe will hardly be enough. Considerations of this kind apply to most farming operations; but whereas some jobs, like haymaking, must be done within a few days or even hours if they are to be successful, others, like sugar-beet harvesting, can safely be spread over a few weeks or months.

Using American nomenclature, the actual average working rate of an implement or machine is called its *Effective Field Capacity*.

The *Field Efficiency*

$$= \frac{\text{Effective Field Capacity}}{\text{Theoretical Field Capacity}} \times 100 \text{ per cent}$$

where *Theoretical Field Capacity* is calculated simply by multiplying the distance travelled in an hour by the effective working width. Field efficiency varies considerably with the type of machine. Typical figures are, for tillage implements, 75–85 per cent; for drilling, 60–70 per cent; fertilizer distribution 40–65 per cent and combine harvesting, 60–75 per cent. Assuming a fairly high field efficiency of 82·5 per cent, the field capacity of a machine in acres per 10-hour day is simply obtained by multiplying the forward speed in miles per hour by the working width in feet. For example, a 7 ft (2 m) cultivator working at $3\frac{1}{2}$

m.p.h. should do $24\frac{1}{2}$ acres (10 ha) in 10 hours or $2\frac{1}{2}$ acres (1 ha) per hour. This rate of work is, of course, only achieved if the work is not interrupted more than usual. It makes no allowance for meal-times or stoppages other than those that may be expected with serviceable equipment and a competent driver. This type of calculation should not be applied to jobs such as spraying, where filling the tank may take as long as the actual application. For such jobs field efficiency may fall to less than 50 per cent.

The effective field capacities of various machines can often be extracted from farm records. Some published figures, such as the results of national machinery testing schemes, have to be used with care.

Most reports include both 'net' or spot working rates, which are equivalent to theoretical field capacity, and 'overall' rates which indicate effective capacity. Often, there are two sets of figures for overall rates, viz. those obtained in testing work under the direct control of test station operators, and those obtained from a survey of the experience of commercial users of similar machines. The figures given in Table A.1 are based on testing where reports are available, and on other published work or experience where they are not. Figures for both spot and overall rates are included, together with forward speeds, width of typical implements and field efficiencies. Where performance in practice falls short of the overall rates quoted, a study of other physical details, in comparison with those quoted in Table A.1, can sometimes help to indicate the main cause. It should be emphasized that the overall figures are intended to represent reasonably efficient performances for the equipment specified. These rates are neither, on the one hand, theoretical and unattainable, nor, on the other hand, averages of current practice which include a high proportion of unnecessarily low figures. As further test results become available, new figures should be substituted for those given in the table.

The final column of Table A.1 gives output per day, and these figures, used in conjunction with information on the possible season of use and the number of working days available, make it easy to calculate potential area per season. The figures per day are based on a standard day of 8 hours. Potential areas on an individual farm are, of course, influenced by several management factors. For example, a tractor and plough which can achieve an overall rate of 1 acre (0·4 ha) per hour can cover a vast acreage on a large farm, where acreage to be done is not a limiting factor, and where there are enough staff to ensure that the plough works about 10 hours a day for weeks on end. The situation on a typical 2-man farm is very different. At the time when ploughing needs to be done there are many other equally important tasks needing attention. This means that though the implement has sufficient potential output to do the work several times over, this potential cannot be taken advantage of on the particular farm.

Capital Invested in Equipment

There is no simple yardstick which determines how much capital can be profitably invested in farm machinery. British farmers spend about £200 million annually on new machinery, and this represents about £6 per acre (£15/ha) of farm land or about £15 per acre (£37/ha) of tillage. In spite of this high annual expenditure on new equipment, average valuations of implements and machinery are only in the region of £20–£30 per acre (£50–£75/ha). It must be remembered, however, that individual figures can differ greatly from averages. If a new £5000 combine is bought on a farm of 500 acres (200 ha) there is likely to be a substantial increase in machinery valuation and, sooner or later, valuations on highly mechanized farms inevitably tend to rise above £25 per acre (£62/ha).

Equipment valuations usually represent only a small fraction of the cost of buying a complete set of new equipment. The cost of doing this would be at least twice the figure quoted on large farms, and nearer four times that quoted on smaller ones. Fortunately for them, few farmers ever have to start from scratch in buying equipment. When this is necessary on a small farm, full mechanization with new equipment is out of the question, since the capital cost would be uneconomic. The only possible way to start farming economically on a small scale is to find a way of sharing the use of equipment, or buying most of it second-hand. A reasonably good set of equipment can usually be bought at less than half the price of new. It is mainly by relying on older machinery that small farms are able to keep implement valuations within reasonable bounds. When equipping larger farms from scratch, it is often worthwhile, where capital is available, to buy new a minimum number of the essentials such as tractors.

In progressive farming areas it is usually possible to buy quite reliable second-hand machines such as combine harvesters and pick-up balers. The older machines may lack some refinements which are very desirable but not essential. Often, however, they have been justifiably discarded by well-established farmers more as an insurance against future mechanical or economic worries, rather than for lack of any particular feature. The financial disadvantages of a high capital

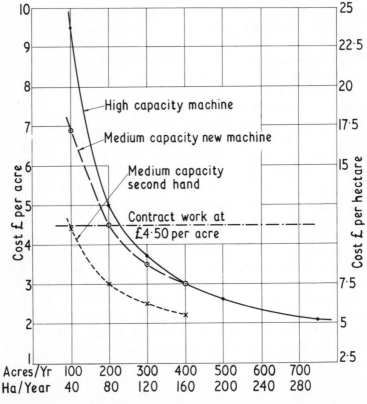

Fig. 2. Combine harvesting costs in relation to area harvested annually.

cost in relation to the amount of work to be done are well illustrated by the problem of choosing a combine harvester. Table 9 and Fig. 2 show typical operating costs for three types of combine harvester in relation to the area worked. Assumptions made are: that the machine of high capacity costs £6000; that a new machine costing £4000 is capable of handling up to about 400 acres (160 ha) per season; and that a machine bought second-hand for £2000 has done about 400 hours work in 2–4 years and is of a type that costs £4000 new. The assumptions made concerning depreciation are based on Table A.2(ii) rather than A.2(i), since this gives a more realistic assessment for the new machines, which are assumed to be replaced in 3–5 years. It should be added that if it is desired to study replacement after only one or two years of use, the appropriate higher depreciation rates shown in Table A.2(ii) should be substituted. Depreciation figures for the second-hand machine are based on the second-hand price; but repair costs for this machine are based on Table A.3 and the *new* cost, with 50 per cent added in order to make some allowance for the machine's greater age and former use.

The calculations show how heavily the fixed charges of the more expensive machines weigh at small usage levels. It might be argued that because cost per unit area of using the high-output combine falls to a reasonable level at about 200 acres (80 ha) harvested, a good case can be made for having such a machine, in view of the advantages that it gives in respect of timeliness in doing the work. However, either of the other two machines is capable of doing the work in good time on such a small area; and it would be a serious waste of capital to use the £6000 machine only to do such a small amount of work. Experience shows that even in a wet harvest there is enough suitable combine harvesting weather in the main corn growing districts to enable 175–200 hours of work to be done on normal cereal cropping. The second-hand machine is, in fact, adequate for up to about 250 acres (100 ha) per season, and it provides the lowest-cost solution for small farms. Choice of a second-hand machine depends on the farmer's abilities and inclinations, the availability of a speedy repair service in case of need and on whether the capital which can be saved on the combine is urgently needed for other purposes.

Table 9. Cost of Combine Harvesting in Relation to Area Harvested Annually

Type and price	High-output £6000						Small/medium £4000				Small/medium Secondhand £2000			
Overall work rate, ac/h	5						2				2			
Area worked annually, ac	100	200	300	400	500	750	100	200	300	400	100	200	300	400
Area worked annually, ha	40	80	120	160	200	300	40	80	120	160	40	80	120	160
Depreciation rate, %*	10	10	11	12	13	15	10	13	15	16½	10	13	15	16½
Annual fixed charges, £														
Depreciation*	600	600	660	720	780	900	400	520	600	660	200	260	300	330
Interest on capital †	240	240	240	240	240	240	160	160	160	160	80	80	80	80
Housing, licence, insurance	36	36	36	36	36	36	24	24	24	24	20	20	20	20
Total annual fixed charges, £	876	876	936	996	1056	1176	584	704	784	844	300	360	400	430
Repairs ‡	50	80	100	120	150	210	60	100	140	180	90	150	210	270
Fuel and oil	12	25	37	50	60	90	15	30	45	60	15	30	45	60
Labour §	16	32	48	64	80	120	40	80	120	160	40	80	120	160
Total annual cost, £	954	1013	1121	1230	1346	1596	699	914	1089	1244	445	620	785	920
Cost per acre, £	9·5	5·1	3·7	3·1	2·7	2·1	7·0	4·6	3·6	3·1	4·4	3·1	2·6	2·3
Cost per hectare, £	23·8	12·7	9·3	7·7	6·7	5·3	17·5	11·5	9·0	7·7	10·9	7·7	6·5	5·7

* Based on Table A.2(ii).
† At 8% p.a. on half capital.
‡ From Table A.3. Secondhand based on *new* price, plus 50%.
§ At £0·80/h.

Circumstances can arise where the lowest-cost solution is not necessarily the best. For example, where labour has been cut to a minimum, a high working rate, which enables harvesting to be completed in a very short time, may be more important than a small difference in operating cost.

Similar relationships between amount of annual use, capital investment and operating cost per unit of use, apply to all farm machines, and provide the main basis for sharing ownership and use of equipment. Increasing use of equipment near the lower end of the range of use makes a very large difference to cost per unit of use, mainly because of the disproportionate incidence of fixed costs. At above about half the potential annual use, however, additional use has relatively little effect on cost per unit of use, though it may still have considerable advantages in such directions as limiting capital investment in machinery, and avoiding the consequences of obsolescence.

Even if investment of capital in new machinery will result in an overall reduction in operating costs, it is not necessarily the right thing to do from the viewpoint of farm management as a whole. The farmer sometimes needs to compare the advantage he would gain from investment in the machine with the gain that might be obtained if he invested the available capital in say extra livestock or fertilizer. As the amount of capital required for mechanization increases, it becomes more than ever necessary to weigh its needs against those of all the other necessities for full and efficient production.

Standards for Checking Mechanization Efficiency

For a given region and type of farm, an indication of efficiency in the use of labour and machinery may be obtained by comparing the performance of an individual farm with that of a group of similar farms. Standard methods of computing both the output of the farm and the cost of running it have been worked out, and the necessary figures can be easily arrived at by using standardized accounting methods based on an account book which is nationally distributed through the National Farmers' Union. Some Provincial Agricultural Economics Departments regularly publish both average and 'premium' or 'high-income' standards for *net output per* £100 *labour* (paid and unpaid) and *net output* (excluding pigs and poultry) *per* £100 *machinery expenses*. An unsatisfactory profit level, accompanied by a low figure of net output per £100 labour, may indicate inadequate mechanization. In such a case, the only remedy on a small farm is likely to be increased output; but on large farms there may be a possibility of achieving higher efficiency by reduction of the labour force.

A low net output from crops and grazing livestock per £100 machinery expenses on an individual farm may indicate that an excessive amount of money has been spent on machinery, or that the equipment is being inefficiently used. In such a case it is necessary to make a thorough study of mechanization management with a view to making the best use of existing equipment, and planning future investment so that the money spent will produce a higher return.

Farmers can check their own efficiency without great difficulty. The items needed are:

(1) Net output   = Gross income (i.e. sales plus or minus changes in valuation) minus purchases of feeding stuffs, seeds and livestock.

(2) Labour cost   = Actual expenditure on wages plus an allowance for the *manual* work done by the farmer.

(3) Machinery cost = Implement and machinery depreciation (= wear and tear allowances), repairs, fuel, lubricants, contract work, and farm share of car expenses.

It is important to compare the figures obtained with those of similar farms, since there is considerable variation not only between types of farm but also between similar types of different size.

Average and premium figures by type and size of farm are available for some regions from A.D.A.S. or University Departments of Agricultural Economics. With the reservation that there are big differences between farm types, normal ratios have been stated as follows:

Total power and machinery cost = Net output divided by 5·2
Total labour cost = Net output divided by 3·5

These figures should only be taken as a general indication of the expected general order of costs. For any serious study of the subject in relation to an individual farm it is necessary to obtain up-to-date figures for farms of similar type and size. If machinery cost figures are appreciably higher than average, it is advisable first to check whether the cause is a low farm output. If it is below average, individual items such as crop yields and output from various forms of livestock must be carefully studied. Unless output is sufficiently high, a farm cannot show an adequate profit.

An alternative method of checking what machine (or labour) costs 'ought to be' for an average farm is via their ratio to total gross margins. For example, the Universities of Bristol and Exeter[9] suggest the following formula:

$$\text{Total power and machinery cost} = £108 + \text{Total gross margins} \times 0 \cdot 204$$

More specific calculations of a similar type are suggested for particular types and sizes of farm.

Checking Labour Requirements

The amount of labour that should in average conditions be needed for any farm can be calculated from a knowledge of the average labour requirements for the production of the individual items, according to the figures given in Table A.4.

Annual labour figures such as those given in Table A.4 have only a limited use. As some of the details indicate, there are in practice wide differences in labour required according to the degree of mechanization; and only more detailed figures related to types of machines and methods of use can give precise requirements in particular conditions. Moreover, the annual figures take no account of the vital problem of seasonal labour peaks. Nevertheless, such annual figures can be useful in making a rapid preliminary check on labour needs.

Comparing Labour Requirements with Labour Available
When the number of man days that should be required has been calculated, the number of days actually provided by the present staff may be reckoned. The total amount of work that the production should normally require (from Table A.4) may be

expressed as a percentage of the number of standard man days available. If the basis used for calculating labour need and labour available chosen is the figures of J. S. Nix,[10] 15 per cent must be added to the estimated labour requirement to allow for unproductive work on maintenance and overheads generally, before the labour efficiency index is calculated.

Labour Efficiency Index
$$= \frac{\text{Estimated Labour Requirement}}{\text{Labour Available}} \times 100 \text{ per cent.}$$

For example, suppose that on a particular farm, the total estimated labour requirement, based on Table A.4 is 720 standard man days, and that the work on this farm is done by a regular staff of 3 men, comprising a cowman, a tractor driver and a general worker.

$$\text{Labour Efficiency Index} = \frac{720}{300 + 250 + 250} \times 100 \text{ per cent}$$

or 90 per cent.

This result indicates that while labour efficiency is by no means perfect, it is not so poor as to suggest that labour efficiency is a major problem on this farm.

The use of labour on an individual farm can be further checked by comparing the cost with that of other farms of a similar type. In doing this it is necessary to consider not only the type of farm but also the size, and the contribution made by family labour. Up-to-date figures taking account of current wage rates and local circumstances may be obtained from most regional Departments of Agricultural Economics.

A generalized formula suggested by the Universities of Bristol and Exeter[9] is as follows:

Total farm labour cost = £544 + Total gross
margins × 0·266.

More specific calculations are suggested for different farming types and sizes.

When estimating the labour requirements for tending livestock it is often useful to consider what is a typical size of stock enterprise for a one-man unit. Table A.5 provides guide figures. As with crop production, mechanization can completely transform the situation, and figures considered good a short time ago are already obsolete. Such questions are further considered in Chapter 10.

Seasonal variations in total labour requirements have been briefly considered on p. 21. The likely effects of changes in cropping and livestock programmes on seasonal labour requirements can be roughly estimated from a study of Table A.6 in conjunction with the total labour figures for crops and stock given in Table A.4. Both the total figures in Table A.4 and the seasonal distribution in Table A.6 are, of course, averages for a wide range of farms, and quite different results may be obtained in some conditions. It should be particularly emphasized that since standards used in different regions may differ considerably, it is advisable when tackling the labour problems of an individual farm to obtain a copy of the appropriate regional standards and to follow these. Standards will naturally be changed from time to time as new developments take place in equipment and methods; so the latest possible information should always be used.

Analysis of Seasonal Labour Peaks

The main factors involved in the incidence of difficult labour peaks are:

Amount (usually area) of each job.
Length of period over which the work can be satisfactorily done.
Rate of work when the job is in progress.
Minimum gang necessary.

To facilitate a study of peak periods on a particular farm, a method sometimes called 'Analysis of Gang work days' was introduced by the Cambridge Farm Economics Branch. The year is divided into a number of seasonal periods, and estimates made, based on average weather, number of daylight hours etc., of the number of working days available in each period. As an example, estimates for the Eastern Region [11] are given in Table A.7. These figures may need modification for other regions, and also for particular conditions, such as for farms on a very heavy soil. It should be assumed that the number of days available will be appreciably less at some seasons in wet areas, especially where the soil is heavy.

The labour requirements of different enterprises can be calculated in detail from the figures given for use of various types of machines in Table A.1, and from information concerning the requirements of particular operations involving gangs of workers, in Table A.8. The other important factor is the period during which the work should be carried out. Some information on this is included in Table A.8.

Armed with this information it is possible to list the operations to be done, the number of worker hours needed and the minimum gang size for the various essential jobs, and to study how these can best be fitted into the number of days available for the work. The task is facilitated by use of specially prepared labour distribution charts, consisting of ruled sheets to assist the drawing of simple histograms depicting the allocation of work days to particular jobs. Planning is assisted by stating machine performance in acres per 8-hour day, as in Table A.1, and in Table A.8. The method of planning may be illustrated by a simple example.

Suppose that on an arable farm, with corn harvest completed by the end of the first week (5th 'available day') in October, the work to be done in autumn is as follows:

Operation	*To be completed by*
Drill 40 acres (16 ha) winter wheat after bastard fallow	End of October
Plough, cultivate and drill 40 acres (16 ha) winter corn after corn	End of November
Harvest 45 acres (18 ha) sugar beet	1st week in December
Plough 80 acres (32 ha) for spring corn	Early January
Livestock work	Continuous at $\frac{1}{2}$ man day

This is a simple programme – somewhat over-simplified compared with actual needs on most farms. Hitherto, a side-delivery sugar beet harvester has been used and this operation alone has required a 3-man gang. A study is to be made of the possibility of using a tanker harvester, and doing the autumn work with only two men.

It is known that the work on sugar beet can be substantially reduced by using the tanker harvester as a one-man machine, the loads being delivered on to a concrete pad. Moreover, the wheat after bastard fallow needs little more than drilling, and ploughing for spring corn can run on into the New Year if necessary.

A detailed assessment of the operations to be done gives the following information.

Operation	Equipment	Gang Size	Rate of Work Acres/day	Acres to be worked	Work Days needed
Plough for wheat	3-furrow plough	1	8	40	5
Prepare land for drilling wheat	Spring-tine cultivator-harrow	1	36	40 once 40 twice }	4
Drilling wheat	Combine drill	2	23	80	$3\frac{1}{2}$
Harvest sugar beet	Tanker harvester	1	$1\frac{1}{2}$*	45	30
Plough for spring corn	3-furrow plough	1	8	80	10
Livestock					$\frac{1}{2}$ daily

* Limited by necessity to transport to pad.

The work days needed may now be fitted in on the labour distribution chart as shown in Fig. 3. The assumptions made are that the sugar beet harvesting goes ahead uninterrupted except by the corn drilling, and that the second man divides his time equally between livestock, and such work as ploughing.

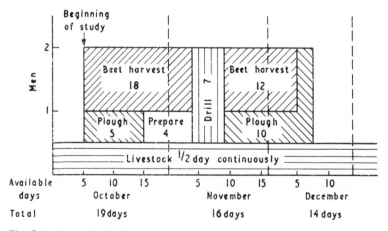

Fig. 3. Labour distribution chart.

The chart indicates that if modern equipment is efficiently used there should be no difficulty in getting through the autumn work with two men. For the sake of simplicity the drilling is shown as one spell in November, but in practice there is no difficulty in drilling the fallow earlier.

Checking Tractor Needs

In the same way as for labour requirements, the average tractor requirements for production of crops and livestock can be calculated on a basis of 'standard tractor hours' introduced by Sturrock. When the conception of standard tractor hours was first established, most tractors were of the size which would now be classed as 'small–medium'. This size of tractor is capable of operating a 3-furrow plough in average soil at a moderate depth at what was then the normal ploughing speed of about $2\frac{1}{2}$ m.p.h. (4 km/h). A standard tractor hour was an hour worked by such a tractor, and suitable adjustments were made for smaller and larger tractors. There is now a much wider range in sizes and outputs of tractors, but the conception of a standard tractor hour is still valid, and average figures for annual requirements for different enterprises are given in Table A.9(i). It should be understood that actual tractor hours will vary considerably, in the case of some enterprises, according to the size of power unit used; and alternative figures are given for farms where high-capacity machines are used.

The differences in ploughing output between a small–medium and a very high-powered tractor can be as high as 4 : 1 in work where the whole of the power of the large machine can be effectively applied; but it is very much less on lighter work, and there are some jobs where the small machine can do more work than the large one. On the average of a year's work for those enterprises where significant differences for different sizes of tractors are shown in Table A.9(i), modern tractors may be 'rated' as in Table A.9(ii). The ratios in Table A.9(ii) should be used only as a general guide for application in conditions where the more powerful tractors can be effectively employed. This will usually be in circumstances where large or very large tractors are not the only power units on the farm.

The calculation of standard tractor hours can be used to check on tractor usage. It was found in the Eastern Counties that a single tractor is usually sufficient where up to 800 hours of tractor work are called for, while additional small–medium powered tractors could each handle up to 1200 hours of work. The total of standard hours as calculated can now be compared both with the number of tractor hours actually worked (if known) and with the number of tractors available. For example, suppose on the crops and stock production basis the answer obtained was 3000 standard tractor hours. One tractor at 800 hours and two at 1200 hours each amount together to a total of 3200 tractor hours, so in theory three small–medium powered tractors should be enough. If the farm is carrying four or five it is necessary to enquire why, and to see whether capital is being unnecessarily locked up in spare tractors. It is, of course, inadvisable to apply the results of such calculations in a mechanical way. Allowance must be made for all unusual factors, and due account must be taken of the great advantage of having a spare tractor in busy periods. Questions such as the relative capacities of tractors of various sizes are further discussed in Chapter 5.

Checking Machinery Costs

When comparing very similar farms, it may be useful to study machinery expenses per acre. This can be misleading if farms of similar size but widely differing intensity of production are compared; but provided such pitfalls are borne in mind, average figures may be compared with those for the individual farm. Regional figures are available for specialized types of farms, such as fen and silt arable farms in the Eastern Counties, where the general level of costs is appreciably higher than the national average for arable farms.

On average, surveys show that, before the 1973–4 increases in fuel prices, about 40 per cent of machinery expenses arose from purchase of machinery or depreciation; about 25 per cent was concerned with repairs, renewals and replacements which are not depreciated; about 20 per cent went on fuel and electricity, 10 per cent on contract services and 5 per cent on vehicle taxes and insurance. There is a deduction of about 5 per cent overall for private use.

It must be appreciated that use of standards can help in diagnosing mechanization deficiencies, but does not, as a general rule, give any clear indication of the remedies that may be applied. If use of the standards shows that mechanization is over-expensive or is not securing adequate economies in the use of labour it is necessary to examine many questions of equipment and method in detail in order to see where appreciable improvements might be made. Some of the points that should be looked for are discussed in Chapters 4–10.

Partial Budgeting

The principles of costing explained in Chapter 2 and applied to calculation of the running costs of combine harvesters earlier in this chapter may also be applied, if due care to avoid pitfalls is exercised, for calculating the effects of many kinds of changes in the use of labour and machinery on individual farms. The most important error to guard against is that of assuming that potential labour savings by the use of machinery can in all circumstances be translated into actual savings in labour costs. While there are many advantages to be gained by saving the time of regular workers, it is usually only if the farm can manage all the year round with one less that an appreciable saving in labour costs is achieved. There is, of course, the possibility of making appreciable savings in overtime, and on some farms there is the possibility that a thorough overhaul of working methods in peak periods can in fact enable the farm to do with one or more fewer workers. An appreciable saving of casual labour is often a good reason for mechanizing a particular job, and here the introduction of a machine can sometimes be an extremely profitable investment.

As an example of the methods of calculation and the principles involved, consider the case of a farmer who grows 40 acres of sugar beet on medium loam. Approximately half of this is thinned by the regular staff, but the rest has to be done by casual labour at a price in the region of £20 per acre (£50/ha). By changing to the use of the most up-to-date drilling and spraying equipment and monogerm seed it is reckoned that the need for casual labour can be wholly removed, without appreciable sacrifice of yield. Cost of the new equipment is assumed to be £600.

A simple budget suffices. The main extra costs are the depreciation, interest on capital and repairs due to ownership of the new equipment. There may possibly be a small extra cost on spray material but it is more likely that a herbicide is already used and that the new equipment will make use more effective.

The work rate (Table A.1(ii)) is taken as 1·1 acres per hour (0·4 ha/hr), so annual use on 40 acres (16 ha) would be only about 36 hours. Table A.2(i) (Group 4) shows a life of 12 years at this level of use; but it is more prudent to assume the average depreciation rate of 10 per cent per annum indicated for light use in the centre column of Table A.2(ii). On this basis, annual fixed costs are:

Depreciation $\dfrac{£600}{10}$ = £60 per annum

Interest, 8% on £300   = £24 per annum

Repairs and spares. Table A.3 2·5% of £600 = £15 per annum

Total extra cost approx.   £99 per annum

This extra cost could be more than halved without any appreciable disadvantage if it were possible to share use with a neighbour. Even if there is some extra spraying cost due to use of more expensive material, the total should still be well below the cost of the thinning casual labour saved. There is, therefore, a strong case for finding the most economic and effective way of securing use of the new equipment.

There is, however, one very important proviso. The new technique must not have a serious adverse effect on crop yield, otherwise any calculated savings can easily be more than wiped out. One ton per acre (1016 kg/ha) reduction in yield would result in a serious reduction of income. Nevertheless, there may be the possibility that a slight reduction of yield will be preferable to having to give up growing the crop.

In other circumstances, an extra mechanized operation is

likely to involve extra tractor work and labour for carrying it out. The cost of such work will need to be borne in mind, but not necessarily charged for in full. In such circumstances the normal procedure is not to charge for tractor fixed costs or regular labour, but merely to make a charge for tractor fuel and repairs.

It should be emphasized that the saving resulting in this case is due to the elimination of casual labour. If the hand work had previously been done by the regular farm staff on piece-work, the saving would have been appreciably less. It would in fact amount to the difference between what the men would earn in the period at normal day rates and the larger amount that they earn on piece rates and with overtime.

Forward Planning of Machinery Replacements

To ensure a reasonable degree of efficiency in mechanization management over a long period it is essential to have a long-term plan for replacement of the major items of equipment. A simple way of preparing such a schedule is to list all of the chief items and to write in details as in the specimen worksheet below.

Item	Model year	Bought year	Cost £	Present value £	Re-maining years	Replacement Year	Cost	Trade in value £	Cash or credit needed £
Tractor	60 h.p. 1973	1973	2000	1400	5	1979	3000	500	2500
Forage harvester	Flail. 1972	1972	600	400	3	1977	900	200	700
Combine drill	20-row 1970	1970	600	300	4	1978	1000	200	800

In planning the replacement year it is advisable to aim to spread replacements fairly evenly. In the event of an emergency due to breakdown or unexpected improvement in a new model, replacement of a particular item can be brought forward. Looking ahead can help in making the right decisions concerning such questions as sharing the use of a particular machine.

CHAPTER FOUR

Work Study and Mechanization

Introduction. Scope and Limitations of Work Study in Farming. Method Study. Multiple Activity Charts. Critical Analysis in Method Study. Use of String Diagrams. Securing Labour Efficiency: Some General Principles: Repetition Handwork: Ergonomics in Agriculture.

Introduction

'Work Study', in its technical sense, is the study of work in a systematic manner, with a view to improving the effectiveness with which the job is done. It embraces 'Method Study' or 'Work Simplification', and study of the time taken for various parts of the task. ('Time Study'.) It may also include a detailed study of movements employed in doing particular jobs. ('Motion Study'.) The methods employed by work study teams to assist management in manufacturing industries have been gradually built up over a period of more than a century into complex techniques concerning which many books have been written, especially in the United States.

It is necessary at the outset to consider broadly to what extent the arbitrary work study procedures which have been developed largely by a process of trial and error in manufacturing industry [12] can be employed with advantage in farming; and also to decide how work study fits in with mechanization, which has precisely the same objectives. Broadly speaking, work study is concerned with the more efficient use of existing resources in men and machines, whereas mechanization generally involves the substitution of mechanical power and equipment for human effort.

Work study has been much longer accepted and utilized in U.S.A. than in Britain, both in industry and in agriculture. In that country the emphasis in agriculture is on 'Method Study' or 'Work Simplification'; and there is no suggestion there that individual work study by professional consultants should be carried out on ordinary farms. Work study is, in fact, considered as a form of research which aims at finding the best ways of doing jobs, in order that these best ways may be recorded and made known to the thousands of farmers who are interested in doing a similar job, possibly in slightly different circumstances. Though farms and working conditions differ, work study or operational research such as that dealing with the use of milking machinery, which is considered in Chapter 10, is applicable with suitable qualifications wherever one or other of the normal types of milking parlour is concerned. Similarly, there are principles dealing with the use of farmyard manure spreaders, forage harvesters and many other machines which can be widely applied without necessarily involving systematic work study on the individual farm. In the case of large farming businesses, especially those with specialized enterprises of a factory nature, there may be advantages to be gained by the employment of a professional work study consultant. As a general rule, however, the way in which small farmers can benefit is in being supplied with the results of researches carried out on both commercial and experimental farms. So far as the most economic use of machinery is concerned there is considerable scope for experimental work, as well as for recording of what occurs in ordinary farming practice.

Motion study techniques are of value in connection with the design of agricultural machines – especially those which involve continuous use of labour, such as hand-fed potato planters, potato harvesters and hop-picking machines. This is an aspect of work study which affects farmers but will not be further considered in this book, since there is little that farmers can do about machinery design, apart from complaining about the more obvious faults.

In the discussions and explanations which follow, some use is made of certain terms which are used by work study specialists. An attempt is made, however, to avoid introducing unnecessary jargon, or adopting a pedantic attitude in regard to terms such as 'Method Study', which should be taken to mean here what they mean in ordinary English usage.

Scope and Limitations of Work Study in Farming

Some eminent work study specialists have declared that work study is, like science, 'largely organized common sense'. This is as good a generalization as any, and it should first be made plain that neither work study nor any branch of it is a substitute for good management or for technical knowledge. Moreover, though its objects are precisely the same as those of mechanization, it is not normally a substitute for mechanization but rather an aid towards ensuring that power and machinery are more effectively utilized. It should be recognized that method study is essentially a study of engineering in its broadest sense; and it is necessary before making decisions on new methods to be sure that account is taken not only of mechanization possibilities that already exist, but also of those which are likely to develop in the future. For example, a detailed study of hand hoeing sugar beet sown by a cup-feed drill might in some circumstances be of value to farmers who cannot, on account of overall management factors, use any other method. Before work study on such a subject is undertaken, however, it is necessary to know that the trend of mechanization developments is towards the use of single-seed drills, with or without mechanical thinning. On many farms an improved drill will do more to reduce labour requirements than will any improvement in the technique of using the old type. It is important, therefore, that full account should be taken of mechanization potentialities before work study on a particular problem is undertaken.

One of the most important features of work study from the farmer's viewpoint is the attitude of mind that it generates. Once he has appreciated what method study involves, there is no job on the farm that he does not examine more critically; and often he can make substantial improvements without any resort even to a wrist watch. The most important step towards securing an improvement is to realize that something is not being done as effectively as it could be; and even a superficial experience of work study techniques can assist a farmer to this point. There is a natural tendency for interest in work study to be greater where profits are low than where they are satisfactory; but since low profits are often associated with a low standard of management generally, it is unlikely that in such circumstances work study can easily secure substantial improvements.

One of the cardinal principles of all work study is the exercise of care in deciding which particular job or aspect of a job should be tackled. This is especially important in farming, where, for example, low profits from milk production are quite as likely to be caused by, say, an unsound general feedingstuffs policy as by inefficiency in say silage making or milking.

Broadly speaking, the benefits from work study are likely to be at a maximum where management is already good, where the labour supply position is tight and where expansion of production is required or is taking place. In these circumstances labour saved can clearly be profitably employed on productive work.

There may be little to be gained, on the other hand, by applying work study in conditions where vital technical factors are the main cause of unsatisfactory results.

One further limitation of work study is that it is no cure for bad labour relations, and can indeed be damaging if clumsily applied. It can, in fact, only do good in conditions where both farmers and workers understand what is being considered, and where the workers welcome a study of their methods. Workers may interpret criticism of working methods as implying criticism of themselves. They may, therefore, resent the results of apparently successful work study. They may, moreover, not welcome a reorganization if it reduces their total earnings by cutting down overtime. It should also be added that a good solution which implies criticism of previous management may be rejected if it is not tactfully presented.

Sufficient has been written to indicate that before work study is seriously considered on an individual farm it is advisable to look into the economic, technical and human factors as carefully as possible. What follows is written primarily for farmers, with a view to assisting in the critical study of work carried out on their own farms.

Method Study

'Method Study' implies, as the term indicates, a study which is aimed as finding better ways of doing a particular job or a number of jobs. This is the branch of work study which is of particular importance and value to farmers. The essence of the work study approach is that the problem should be tackled in a thoroughly systematic manner, the chief stages of the process being (1) to record accurately the method at present in use; (2) to analyse the task in a thoroughly critical way and finally (3) to suggest or decide on improvements. A few simple charts can with advantage be used to assist in the study, and the first of these that is considered is called a 'Multiple Activity Chart'.

Multiple Activity Charts

Multiple Activity Time Charts may be used when it is necessary to consider team work in which several distinct operations are involved, as, for example, in silage making or farmyard manure spreading. Here several tractors and men may be engaged, and there is a regular cycle of operations involving loading, transport and unloading. As an example, silage making using a forage harvester may be considered. The operations involved are (1) drive trailer to field, (2) load trailer, (3) drive loaded trailer to farm, (4) empty trailer, and (5) spread the silage. In studying such an operation, a record must be made of the time taken for each of these individual components of the task, and also of any delays that occur in between. The work in the field, on transport and at the silo must usually be studied in succession, and as there are always variations in the timings for different loads it is advisable to take the average of several. A wrist watch is quite accurate enough for the timing. When the necessary times have been recorded, the information can be charted as in Fig. 4, which shows the operation as it was recorded on a large farm in the Eastern Counties. Each bar of

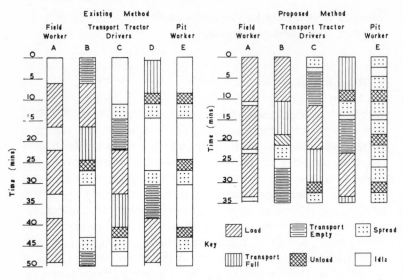

Fig. 4. Multiple activity charts of silage making.
Left: Before work study. *Right:* One possible alternative system.

the chart represents the activities of one of the five workers on the time scale shown at the left. It is important to fit the activities into the appropriate places on the time scale. The forage harvester in this case delivered the crop into a trailer which was pulled alongside by a separate tractor. There were only two tractors and trailers used for transport, and drivers B, C and D took it in turns to drive the empty trailer to the field and collect a load. Three men were engaged at the silo during unloading, viz. the pit worker E, the driver who brought in the load, and the driver who would next take the empty trailer to the field. The important operational times were as follows: loading, 11 min; transport full, 8 min; unloading, 3 min; spreading, 3 min; and transport empty, 8 min. This was not a good system, and the chart shows a high proportion of waiting for all the workers, including the driver of the forage harvester. The farmer complained that the forage harvester was not fast enough, but in fact he was not using the machine to anything like its full capacity. One possible method of reorganization would have been a change to towing the trailers behind the harvester and cutting down the number of men and tractors engaged. This possibility is not considered here, but Fig. 4 shows a reorganization using the same gang and one extra tractor and trailer. Instead of the 3-load cycle taking 50 minutes it should now take about 35 minutes. The chief effect of the change is to cut out the idle time and enable the harvester to keep going. The only worker whose job is appreciably harder is the one who is permanently at the pit. He will possibly need a little more rest than is shown on the chart, and could probably take it at the expense of a little spreading, since it appeared that an unnecessary amount was being done under the existing system. It might also be possible to arrange to distribute some of the heavy manual work among the transport drivers by arranging for one or more in turn to relieve the pit worker. Even without the extra tractor and trailer, the cycle times could have been appreciably reduced simply by sending the trailers on their way as soon as they were emptied. The position was, however, that the driver of the forage harvester was adjusting his speed to that at which the trailers arrived; and neither he nor the trailer drivers realized that each was slowing down the other. An increase in transport speed by improvement of the farm roads would also have had a marked beneficial effect. Such inefficiency is a common feature of unbalanced team work, and one which simple work study can help to remedy. This type of method study is not beyond the capacity of the farmer himself, once he has mastered the technique of recording and charting his observations. His chief difficulty is a psychological one: he invariably finds it extremely hard to view the job effectively, and all too often begins by assuming that there is little wrong with his existing organization. If he can overcome such prejudices, however, there is nothing to prevent his trying his hand at method study. Once he knows the vital times for loading, transport and unloading, he can experiment on paper with various combinations of men and machines, and this will often assist greatly in arriving at a sound system.

Critical Analysis in Method Study

Many farming operations, including some of the jobs about the buildings which are connected with tending livestock, can be profitably subjected to one of the basic method study procedures called the critical analysis. As in so much of the work study routine, the questions asked may at first sight seem childish; but asking and answering them about every little detail sometimes helps to make clear what are the important features, and which parts of a task are unnecessary. For example, consider, in the example referred to above, the spreading of the silage, which was the chief cause of inefficiency in the silage making. The questions that need to be asked are as follows:

(1) (*a*) WHAT is achieved? (*b*) Why is it necessary? This raises the question as to how much spreading is really needed when ensiling in a pit silo, green crops that have been cut and lacerated by the forage harvester.

(2) (*a*) WHERE is it done? (*b*) Why there? This particular question is irrelevant in this case, but is an important one in the case of many jobs such as weighing produce, which may be done in a number of different places.

(3) (*a*) WHEN is it done? (*b*) Why then? With the silage job, spreading could possibly be done other than after every load is emptied. In an extreme case it might be possible to do the job at the end of the day, *and such unlikely ideas should not be entirely excluded from consideration.*

(4) (*a*) WHO does the job? (*b*) Why that person? This question also is an important one in respect of the silage spreading. Is it really necessary to have a permanent pit worker to do the job? Could the transport drivers do it?

(5) (*a*) HOW is it done? (*b*) Why that way? This is another vital question. The silage is probably spread by hand fork. Must it be done that way when a 35–40 h.p. diesel tractor is probably standing ticking over nearby? Is there no way of using the tractor to do the job?

If such an analysis is applied to the various elements of a job, the answers should show where change is most likely to be effective, and where improvements are possible.

The critical examination does not, however, end with the 'Why?' questions. The development of alternative methods may, in fact, be assisted by asking questions on all the possible *alternative* ways of doing the job, followed by questions as to what SHOULD be done. For example, the questions 'Who does it?' and 'Why?' are followed by 'Who *else* could do it?' and finally 'Who *should* do it?' To summarize, the questioning pattern is as follows:

	Challenge	Alternatives	Pointers
Purpose	What and why	What else	What should
Place	Where and why	Where else	Where should
Sequence	When and why	When else	When should
Person	Who and why	Who else	Who should
Means	How and why	How else	How should

Finally, decisions or at least suggestions must be made, and at this stage it is necessary to consider (1) Purpose: the possibility of eliminating the particular job as an unnecessary one; (2) Place, Sequence, Person: combinations and changes to improve the method; and (3) Means: the possibility of simplification of the way of doing the job. E.g. in the silage spreading, how useful is a tractor-mounted silage spreading rake?

This general pattern of questioning, which is one adopted by industrial work study specialists, can with advantage be adopted in carrying out a study of many farming techniques.

Putting Findings into Practice

The industrial work study specialist does not leave a job at the point where the management decides to introduce a new method: on the contrary, he supervises the inauguration of the improved method, and returns after a time to make sure that it is working as well as possible. In the same way, a farmer who adopts a new technique must make sure that it is properly applied, both at the outset and as a matter of routine. A system that is fundamentally sound can fail completely if insufficient attention is paid to ensuring that the operators understand what is required of them. They must be shown in detail how best to carry out their duties.

Use of String and Flow Diagrams

A string or flow diagram is sometimes of value for studying jobs such as milking and stock feeding, where the work is done within a limited area and consists partly of irregular journeys between a number of fixed points. In order to prepare a string diagram of a particular task, a scale layout drawing of the working area must first be prepared, and on this must be marked the walls, doorways, machines and storage places. The completed drawing is then fixed to a board and pins driven in at each terminal point of movement and also at each place where direction is changed, as in going round a machine or obstruction, or through a door. A strong thread is then attached to the starting-point pin, and the string is passed in turn round the various points in the order in which the operator moves about. A flow diagram is similarly prepared with the help of coloured pencils in place of string.

As a general rule, the string diagram is made after the information needed has been compiled on a special form which gives the time at which each movement occurred, the destination, and notes on any particular circumstance. In some cases, however, it is possible to make up the diagram as the work proceeds. The completed diagram is merely a special kind of record which helps to bring out excessive or unnecessary movements caused either by unsuitable layouts or poor operating procedures. It enables distances travelled when employing various operating procedures to be easily calculated by relating the amount of thread used to the scale of the drawing. A certain amount of experimental work can be done on the likely effects of moving various items and adopting new work routines. When a good solution of a problem involving a new routine has been found, the string diagram may provide a good means of putting over the findings to the operator concerned.

Securing Labour Efficiency: Some General Principles

It is not necessary to carry out a systematic work study investigation in order to detect some of the common examples of inefficiency that may be found on farms, and a list is given

below of some do's and dont's which may help farmers to see
their own weaknesses. Some of the points mentioned are
elementary, but no apology is necessary for drawing attention
to them, since though they are widely accepted, there are many
farms where labour is continually wasted simply because the
farmer has not questioned an old-established procedure. Farmers are advised to accept the challenge in this enunciation of
general principles by asking themselves whether their own
practices are efficient in each of the particular details mentioned. In some cases an improvement merely needs a decision
by the farmer. In other cases, where the landlord's equipment is
involved, a good solution may be more difficult to achieve.
There is considerable overlapping between several of the points
as set out, but it is thought that this may assist in helping
farmers to pinpoint inefficiencies.

A. **Work about the Buildings**

(1) *Site new buildings or re-site enterprises to reduce or
eliminate unnecessary transport.*
The opportunity for siting or re-siting buildings does not often
arise, but farmers can sometimes make advantageous changes
in the use of existing structures. Many farmsteads have never
been planned as a whole, but have simply grown by the addition of one building at a time. In some cases the needs of today
are different from what they were when the building was constructed. In others new buildings have recently been added
without much thought for materials handling on the farm as a
whole. For example, where a new bulk grain store and a new
piggery have been built these are sometimes found a considerable distance from one another and from the old granary where
the home grown corn is ground and the food mixed. This
means that grain which is efficiently handled in bulk by the
combine harvester and never touched by hand while in store
has to be laboriously sacked off and loaded into a trailer for
transport to the granary. In due course there may be a further
sacking up of the mixed meal, and probably transport in a
trailer to the pig fattening house. It is usually a simple and very
desirable step to eliminate one of these entirely useless sacking

and transport operations by moving the food preparing operation to, or adjacent to the bulk grain store. There may be technical problems involved, but they are usually not insuperable ones.

(2) *Plan stock buildings and yards to reduce work on feeding and littering.*
If milking of cows and collecting of eggs are excluded, feeding bulky foods, cleaning up manure and littering with straw together occupy about 80 per cent of the work required for tending livestock. (See Chapter 10.) In order to reduce this labour it is necessary to plan installations so as to avoid adding unnecessarily to the labour needed. For example, many studies have shown that cattle drop a high proportion of their dung and urine immediately where they feed. This means that where reduction of labour on providing bedding is desired, nothing could be worse than an old-style Eastern Counties bullock yard, with the manger at the back of the covered part. Here, as a general rule, everything is almost as wrong as it could be, both as regards saving of labour and comfort of the stock. The mangers are often completely inaccessible from outside; most of the manure is dropped where the cattle should rest; and the straw needed is often in a Dutch barn so far away that use of a tractor and trailer are necessary to move it. There is little difference between this point and No. 5.

(3) *Plan fixed work areas within buildings to reduce transport, and eliminate unnecessary movement.*
Examples include arranging dairy equipment and milking routines to cut down unnecessary carrying of milk and walking; and handy location of food supplies in a piggery. Again, in a deep-litter laying house, placing the nest boxes where they can be easily emptied saves a great deal of time and trouble.

(4) *Arrange for materials to flow in one direction.*
A plant for grinding and mixing stock feed is an example of an installation that calls for care in planning, so as to have the bought-in foods join the home ground meal at the appropriate place, and the mixed rations finally stored near the door where they will be dispatched.

(5) *Eliminate unnecessary hand work.*
One of the most common ways of wasting labour on farms is allowing men to walk to and fro time after time and day after day with fork-fulls and bucket-fulls of food, and fork-fulls of litter or even manure. Often men may be seen battling their way through the stock to reach inaccessible troughs or mangers. This is a sheer waste of time, and indicates a low level of labour efficiency. Much can often be done by re-siting troughs and mangers to places where they are easily accessible to wheeled transport, or by opening up yards so that a tractor and trailer can go inside easily when bulky fodder or litter is needed.

Sometimes it is possible to stack baled straw so that it can be rolled straight into the yard when needed. It is also often possible to solve feeding problems by arranging for bulk food hoppers, self-feeding of silage, and so on.

The need to make the fullest use of piped water supplies is now widely appreciated, but it is still all too common to find men carrying pails of water day after day in situations where a very little extra spent on pipes and taps, or even a hose, would make the job unnecessary.

Several detailed recommendations arise partly out of the foregoing. They include:

(6) *Use wheeled transport to eliminate unnecessary carrying of materials by hand.*
A good example of the application of this principle is use of a rubber-tyred feed barrow instead of buckets in a piggery. Many journeys are thereby saved and the work is made much easier. Similarly, feed trucks can with advantage be used to assist feeding silage in cowsheds.

(7) *Improve paths, alleys and doorways to suit wheeled transport.*
This follows from the previous item. For successful use of simple transport trucks it is necessary to make paths reasonably level, wide and smooth. This is easily done by the farmer himself, by concreting.

(8) *Make full use of the force of gravity.*
Now that mechanical equipment for lifting heavy loads is commonly found on farms in the form of hydraulic front loaders on tractors, and similar powerful devices, use can often be made of the force of gravity even on farms where the buildings are on a level site. For example, bulk food hoppers above the milking parlour enable the food to be delivered by an enclosed chute directly into the parlour. The use of roller conveyors in packing sheds and light portable conveyors can be extremely useful on many horticultural holdings.

(9) *Make use of cheap and efficient mechanical devices to simplify work or eliminate it.*
One of the best examples is the ball valve control for water troughs, but there are many others which farmers need to consider from time to time. For example, the milk pipe line which eliminates the job of carrying milk to the dairy is one of the important factors in the efficiency of milking parlours; the cattle grid on some farms saves endless time in opening and closing gates. Gates themselves can be effective or a nuisance. One item of fixed equipment which can be most useful in assisting transport on large arable farms is a concrete apron sited where it can be used for storing manure or heaps of sugar beet. In some areas the apron may also serve for storing and possibly self-feeding of silage. Such loading areas may become even more valuable if systems of handling crops in simple containers are developed. The usefulness of such an apron for handling sugar-beet and manure is easily understood. Both materials are easily handled by a front loader working on a hard surface, whereas either job can become almost impossible in the worst of wet conditions. Siting of the apron in the most useful place is worth a good deal of thought. It must be served by a hard road, and easily accessible to as large a number of fields as possible.

(10) *Provide suitable hand tools and locate them where needed.*
This is an obvious need, but the suitability aspect needs some thought. In keeping a 2-level milking parlour clean, the handiest and quickest way is often not with a brush or hose,

but with a bucket and open tank of water, located at the upper end of the pit.

B. Field Work and General

(11) *Employ operating techniques that reduce transport times and idle running, and eliminate unnecessary hand work.*

In all mechanized operations there are efficient and inefficient ways of doing the job, quite apart from such points as correct adjustment of the working parts. For example, the simple operation of ploughing a field can be carried out either by an efficient 'systematic ploughing' method whereby idle headland running is reduced to a minimum, or by a haphazard method which may produce a satisfactory final result but takes longer and uses more fuel. In some other jobs the best operating techniques have not yet been fully worked out. For example, there are differences of opinion on the best ways of using forage harvesters and combine harvesters, and there have been few experiments aimed at providing the necessary data to enable sound judgments to be formed. The need for studies is demonstrated by the simple illustration in Fig. 5. A and B represent plans of a square field which is to be given a uniform dressing of farmyard manure using a trailed type mechanical spreader. There are many ways of arranging the spreading, and A and B merely represent two possibilities. For the sake of simplicity it is assumed that the spreader load is distributed in a distance equal to the length of one side. A common method of working is shown in B, where the field is worked from one of the far sides. A shows a method of working which cuts out a considerable amount of idle running, and the calculation shows how, in a particular part of the field near the gate, dividing the field into two separately worked parts actually halves the idle running time. A few other operational problems of a similar nature are noted in subsequent chapters, but as an example, mention may be made here of the necessity for drilling sugar beet fields and planting potato fields in a manner to facilitate harvesting. In both cases a strip 8–10 yards (7–9 m) wide should be planted or sown parallel to the field boundary all

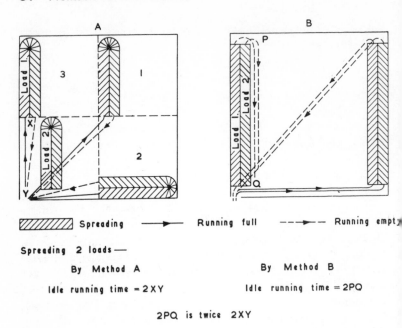

Spreading 2 loads —

By Method A	By Method B
Idle running time $= 2XY$	Idle running time $= 2PQ$

2PQ is twice 2XY

Fig. 5. Alternative methods of using a trailer-type farmyard manure spreader in the field.

round the edges of the field, and any unplanted headland that is necessary should be between these rows and the rest of the field. At harvest time the headland rows are lifted first, and this provides a very wide headland for machines and transport to turn on.

(12) *Haul as large loads as practicable.*
This is merely an extension of the previous principle. As a general rule, large loads save time compared with small, but not necessarily so if much hand work is involved in loading and unloading. Indeed, if a small load can be put on and unloaded mechanically, this will in many circumstances be better than putting on large loads by hand. Thus, a buckrake is often preferable to a trailer for handling long green crops, but a buckrake carrying a $\frac{1}{2}$ ton (508 kg) load is better than one which takes much less. In hay and straw handling, increased load size is one of the reasons for baling. Chopped hay and straw would be more attractive if larger loads could be carried.

Large loads also greatly increase technical efficiency in carting and spreading farmyard and liquid manures.

(13) *Use bulk handling techniques where suitable.*
Nobody who has had experience of handling grain in bulk would wish to use sacks for the job, and bulk handling is similarly attractive for carting potatoes and other root crops. The handling of hay and straw is an exception to the general rule at present, partly because compression of the crop into bales reduces the size of a load and partly because it makes packages which are easily separated from the mass. It seems likely that in due course materials such as high quality hay will in fact be wafered rather than baled, in order further to reduce the storage space, to facilitate bulk handling and assist in preventing waste. With modern forage harvesters, tower silos and mechanical feeders, fodder crops can be harvested, transported, stored and fed to stock without ever being touched by hand. The later stages of such handling systems are also well suited to a considerable degree of automation.

The use of stillages is well worth considering where loads of regular size need to be prepared at a particular point and moved to another place for storage or transport. A modern tractor fitted with a hydraulically operated fork-lift can efficiently move and stack quite heavy loads, and it seems likely that the stillage method can be greatly developed with advantage on many farms for handling feedingstuffs, fertilizers and crops of various kinds.

(14) *Use mechanical equipment to eliminate unnecessary jobs where the capital and operating costs can be justified.*
The reader will appreciate that this principle is a very broad one, covering the whole of the subject matter of this book. The whole basis of successful farm mechanization is, in fact, the elimination of unnecessary hand work and the easing of such as cannot be eliminated. It would be superfluous to quote individual examples, and all that need be said is that there are a few jobs such as potato picking on heavy or very stony land which have so far not been successfully eliminated, but that there are few hand jobs in ordinary farming where elimination,

rather than improving the technique of hand work, is not a reasonable objective.

(15) *Before buying new machinery, first decide the method of operation that is to be employed.*
This is one of the most important rules for successful, i.e. economic mechanization. Examples are given in subsequent chapters of the varied ways in which individual machines may be utilized. For example, a flail type forage harvester may be used to cut and load direct or to pick up from a swath or windrow: some machines can be used to deliver the crops only to a trailer drawn behind, while others will deliver either to the rear or to a trailer running alongside; and many can be used by one man alone, by a team of two, or by a larger gang. Unless a farmer goes thoroughly into all these possibilities and decides what he needs he is quite likely to buy the wrong machine. Moreover, it will be seen in the case of work such as farmyard manure handling how a decision to work over a longer period with a small gang can limit the capital required for mechanization.

(16) *Where there is a choice between using a large gang and a small one, the smaller is likely to be more efficient.*
One of the most significant developments in farm mechanization has been the emergence of efficient methods which require only small gangs for successful operation. Examples include manure loading and spreading by a single man using a hydraulic loader and an easily hitched trailer spreader; one-man silage making using a forage harvester, and a 2-man gang for potato planting, with a mechanical potato planter. Impartial investigations show that with the larger gangs there is always difficulty in avoiding having one member of the team hold up the work of another, and the effect is that a large gang using a great deal of expensive equipment is often less efficient than a small one.

(17) *Remove obstacles to cultivation.*
There are many farms which are unnecessarily difficult to cultivate on account of the shape and size of the fields, bad field entrances, overgrown hedges, and odd trees which have ceased

to be either useful or pleasing to the eye. Now that ordinary farm tractors can be equipped with small dozer blades and a hydraulic loader it is often easy for farmers to improve bad conditions in slack times by using their own equipment. Alternatively, a small bull-dozer can often be hired at a reasonable cost to do a particular job. In this way, field boundaries can often be removed or straightened, with considerable advantage to future cultivation. Where open ditches are filled in, care must, of course, be taken to provide for effective drainage as necessary.

Repetition Hand Work

One branch of work study is concerned with the details of hand work on jobs in which little or no equipment is used. For example, there is advantage in studying such principles as the need to use both hands on work like potato picking. This particular branch of work study is not, however, pursued in this book, partly because an increasing number of hand operations in farming can be more economically carried out with the help of mechanical equipment.

Ergonomics in Agriculture

Advances in mechanization have completely removed much of the heavy toil from farm work; but in a few instances mechanization has introduced new tasks which impose strains on the operators. Examples include the effects of working in a dusty or fume-laden atmosphere; riding for long hours on tractors which are fitted with only a poorly designed seat; and working in a badly constructed milking parlour. 'Ergonomics' is the term given to a study of work as it affects the operator. Detection of strain will often involve tests of a medical nature; but the solution of a difficult problem almost invariably lies in better engineering design. This aspect of mechanization tends to assume increasing importance as living standards rise. Though advances in practice depend largely on improved engineering design, farmers need to bear in mind ergonomic aspects when planning installations of fixed equipment.

Finally, it may be re-emphasized that there is no need for individual farmers to wait for a work study expert to advise on many aspects of improving their efficiency. Many of the labour saving principles mentioned are so clear and easily understood that the farmer need not contemplate systematic work study before introducing a necessary change. In other cases, the best techniques of using new types of equipment are not yet fully worked out, and individual farmers who wish to get the best out of men and machines need not be afraid to do a little investigating on their own, along the lines already explained. Some of the information available on use of particular types of equipment, and a few of the problems still awaiting solution, are further considered in the following chapters. In some cases such as the design and use of milking parlours the subject has been thoroughly studied and the best solutions are reasonably clear. In others, little work has so far been done, and the best ways of working must for the present be based on general experience rather than on systematic work study.

CHAPTER FIVE

Farm Power

Power for Field Work. Types and Numbers of Tractors Needed. Tractor Utilization. Annual Use and Cost of Operation. Operating Speeds and Type of Transmission; Optional Fittings. Tractor Types. High Powered Wheeled Tractors. Efficient Tractor Operation. Power for Stationary Work.

Power for Field Work

The transition from horses to tractors as the source of power for field work in Britain was rapid. The numbers of working horses reached a peak in the period 1906–10, when there were nearly a million in England and Wales alone. The striking changes since then are indicated by the figures in Table 10.

In the whole period over which work horses were replaced

Table 10. Numbers of Farm Tractors and of Working Horses used for Agricultural Purposes in England and Wales

Period	Work Horses	Farm Tractors over 10 h.p.
1871–75	838 000	nil
1906–10	968 000	nil
1921–25	796 000	about 2 000
1939	549 000	55 000
1950	289 000	259 000
1958	73 000	379 000
1964	not recorded	389 000
1972	not recorded	370 000

by tractors, almost 1 million h.p. in the form of working horses was replaced by about 15 million h.p. in the form of tractors of over 10 h.p., plus a considerable amount of power for field work in the form of smaller tractors (44 000 under 10 h.p. in England and Wales in 1972), and many thousands of powerful self-propelled machines such as combine harvesters. The amount of tractor power available increases as tractors become more powerful, while total numbers remain high. The total amount of power used in agriculture also includes a steadily increasing amount in the form of stationary electric motors and portable small engines.

There is little point in trying to estimate accurately the increase in effective power available for farming. It is certainly not less than 15 times as great today as it was half a century ago, and if calculated on a straightforward horse-power basis the ratio would be nearer 20 : 1. But one of the characteristic features of the development of use of mechanical power on farms is that as the amount of power available increases there is a tendency for specialization to develop, and in consequence, average annual utilization of power units tends to fall. By 1957 there was, on average, a tractor of over 10 h.p. to every 67 acres (27 ha) of crops and grass or to every 37 acres (15 ha) of arable land in England and Wales.

Over the period since 1910, the saving in land needed to feed the working horses displaced, reckoned at 3 acres (1·2 ha) per head, must be more than $2\frac{1}{2}$ million acres (1 million ha); and there can be little doubt that this land is better employed in growing food for productive stock or for direct human consumption.

Horses are now seldom used because modern mechanical power units are cheaper in capital cost per unit of power, much easier to maintain and much cheaper to operate because of the extra power in the hands of the worker. Though horses are still used in less developed countries, there is no point in discussing their use on British farms.

Types and Numbers of Tractors Needed

It has been briefly explained in Chapter 3 how the number of hours of tractor work needed can be roughly estimated from a knowledge of the production of crops and livestock, and it has been noted that a second tractor is normally needed when the estimated requirement exceeds 800 hours of tractor work, whereas subsequent tractors can be expected to look after an additional 1200 hours of work each. These figures can only be a rough guide, since there are many important differences between farms, as well as between tractors. An example of one of the differences between farms, not taken fully into account in calculation of the estimated requirements, is the effect of soil type on the speed at which work can be done. If we consider an operation such as ploughing for potatoes, or drilling wheat, a tractor of a given size will usually work appreciably faster on light land than on heavy. As a general rule, there will be a difference of a furrow width in the plough used; and similarly, the drill may be an 8 ft ($2\frac{1}{2}$ m) model in the one case and a 10 ft (3 m) in the other. Alternatively, a higher gear may be possible on a lighter soil. Similarly, it would be reasonable to expect some difference between tractor needs for an acre of sugar beet grown on a deep, heavy silt which is ploughed to a great depth and needs a good deal of working to prepare a seed bed, and those of a similar crop grown on a coarse sand where cultivation is not so deep and seed beds are easily prepared. At the other end of the season, harvesting a heavy crop of laid corn on the silt obviously takes longer than a very light standing crop on the sand. Investigations so far carried out indicate, however, that if mechanization is looked at as a whole, as in making comparisons of total machinery expenses per 100 standard tractor hours, these differences are not as large as might be expected. Nevertheless, when considering tractor needs only, it would be unwise to ignore these differences in power requirement, the extremes of which can be easily demonstrated in practice. On a farm where the land is known to need more than the average amount of power, on account of soil type, working depth or number of operations needed, it will be advisable to reckon on a little more tractor power being needed for each acre of particular crops. If each crop is considered individually throughout the whole year, the mistake of making unnecessarily large adjustments will be avoided. Further research will be needed before more detailed standards can be prepared.

Passing on to a consideration of the numbers of tractors

needed, there is a further effect of soil type that needs to be taken account of, viz. its influence on the seasonal distribution of tractor work. There are some heavy-land farms where, in a wet autumn and winter, the time available for cultivations is very limited compared with that of a farm growing similar crops on lighter, free-draining soils. In the most difficult conditions it is inadvisable to reckon that a 'standard' tractor can usefully work 800 hours (first tractor) or 1200 hours (subsequent tractors) per year. Assuming that the type of tractor is the same, and the average figures for 'subsequent' tractors is 1200 hours, it seems reasonable to reckon on up to 1400 hours on a light, mixed farm with important livestock enterprises, and no more than 1000 hours on a very wet and heavy arable farm with little stock.

Tractor Sizes

One of the difficulties that attends all economic studies of tractor usage is the fact that a tractor is a very variable item of equipment. Engine power, which was predominantly 25–30 h.p. in the early 1950s, doubled to 50–60 h.p. in the next 20 years and continues to rise. When making comparisons of tractor power, it is necessary to ensure that the figures from comparable test methods are used, since there are big differences between different standard test procedures. The main types of standard test include:

(1) A British Standard test for bare engines. This is designed to measure basic potential. It is unrealistic as a tractor test since it does not allow for the considerable amounts of power needed to operate essential engine accessories such as the cooling fan.

(2) Net engine power to I.S.O. tractor test code. This provides a realistic measure of net engine power in that all essential engine accessories must be fitted; but it does not allow for driving essential tractor accessories such as the pump for hydraulic control system.

(3) p.t.o. power to I.S.O. (or O.E.C.D.) tractor test code. This is the most useful standard in that it measures net power available for driving agricultural machines, with

all the necessary tractor accessories in operation at the minimum power absorption level.

A 'medium' tractor is typically in the region of 50 h.p., and a large–medium one not less than 60 h.p. In addition to large–medium mass-produced tractors with engines of about 65 h.p. there are a number of large ones with engines of 80 h.p. or more.

The classification shown in Table 11 is an arbitrary one, but it serves to emphasize how wide a variation there can be in the

Table 11. Tractor Types, Sizes and Typical Prices

Main type	Power group	Approx. power range, h.p.	kW	Typical capital cost, £ per tractor	per h.p.
Market Garden	Motor hoes	1–3	0·75–2·2	100	40
	2-wheel general-purpose	3–10	2·2–7·5	300	40
Wheeled, rear-wheel drive	Small	15–30	11–22	1200	48
	Small–medium	31–45	23–34	1500	38
	Medium	46–60	35–45	1700	33
	Large–medium	61–80	45–60	2200	33
	Large	81–100	60–75	3200	35
	Very large	Over 100	Over 75	5000+	42
Wheeled, four-wheel drive	Large–medium	61–80	45–60	3200	44
	Large	81–100	60–75	4300	44
	Very large	Over 100	Over 75	5000+	42
Tracklayers	Medium agricultural with linkage	70–90	52–67	5500	67

results and cost of one hour's tractor work. When attempting to decide capacity of a tractor for heavy work, a useful guide is the number of plough furrows of normal depth that can be pulled in medium land. Typical figures for the wheeled tractor group are:

small tractors	– 1–2 furrows	
small–medium	„ – 2–3	„
medium	„ – 3–4	„
large–medium	„ – 4–5	„
large	„ – 5–6	„
very large	„ – 6–8	„

In calculating how much work various types and sizes of tractors can do, however, it must be remembered that it is only at heavy work that the more powerful tractors have much advantage over smaller ones. Thus, when the number of standard tractor hours available from a particular tractor fleet is being estimated, due allowance must be made for the relative inefficiency of large tractors when doing light work such as fertilizer distribution or swath turning. This has been done in estimating the 'rating' in terms of standard tractor hours in Table A.9(ii).

Tractor Utilization

Studies of tractor use show the following general features:

Average annual use. If 'spare' tractors or old tractors set aside for special duties are excluded, average use of the main tractor fleet is in the region of 700–800 hours annually. Many tractors do less than 700 hours, and a few do as much as 2500. As the average power and cost of tractors increases, there is no general tendency for higher average annual utilization. The newest and most powerful tractors tend to be used a little more than average, but typical figures for this group are only 800 to 1000 hours annually.

Average use is low on small farms and tends to increase up to a farm size of about 500 acres (200 ha). Many very small farms have two or more tractors and there can be sound reasons for this, provided that the capital value of the tractors bears a reasonable relation to the amount of work done. Having more than one tractor can help in securing efficient use of labour in many operations involving transport and handling.

Tractor needs, both annual and seasonal, are greatly influenced by the husbandry needs of individual farms: and the possibility of securing high average utilization depends to a great extent on the soil type and farming system. Average use is considerably higher on mixed light-land farms than on heavy-land arable farms. In some circumstances where annual use is low, as on mainly cereal farms on heavy land, there is full justification for use of higher-powered tractors to get essential

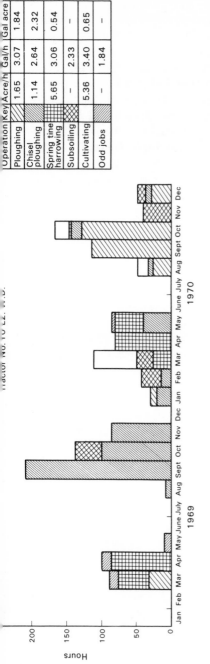

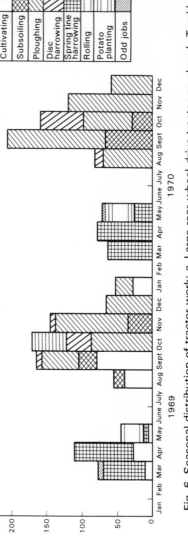

Operation	Key	Acre/h	Gal/h	Gal acre
Ploughing		1.65	3.07	1.84
Chisel ploughing		1.14	2.64	2.32
Spring tine harrowing		5.65	3.06	0.54
Subsoiling		–	2.33	–
Cultivating		5.36	3.40	0.65
Odd jobs		–	1.84	–

Operation	Key	Mean		
		Acre/h	Gal/h	Gal/acre
Cultivating		2.30	2.24	0.98
Subsoiling		1.51	2.85	1.89
Ploughing		1.12	2.53	2.25
Disc harrowing		2.54	2.42	0.95
Spring tine harrowing		4.37	2.79	0.64
Rolling		3.75	2.15	0.57
Potato planting		1.32	1.41	1.07
Odd jobs		–	0.5	–

Fig. 6. Seasonal distribution of tractor work; *a.* Large rear-wheel-drive tractor on clay; *b.* Tracklayer on clay/chalk.
Source: A.D.A.S.

field work done in the limited time during which soil conditions are favourable.

Seasonal pattern of use. The seasonal pattern of use of tractors generally shows two main peaks of tractor work, one in autumn and the other in spring. On heavy land, where a high proportion of crops tend to be autumn drilled, the amount of autumn work greatly exceeds the spring work. The main cultivation tractors need to be chosen to suit the farming needs. Thus, rear-wheel-drive tractors are adequate for the needs of most light-land farms and may be particularly suitable where summer work such as forage harvesting is important. On heavy land it is sometimes essential to continue cultivation when soil conditions are not ideal. In such conditions four-wheel-drive tractors and tracklayers can be effectively used after real-wheel-drive tractors have to stop work. There is sometimes a conflict between a high level of utilization and efficiency in carrying out particular operations, and the best compromise has to be sought.

Figure 6b shows the seasonal work pattern of a tracklayer on a large clay/chalk farm where it could effectively continue ploughing during the winter months. In contrast to this, Fig. 6a shows the seasonal pattern on a soil described as clay with

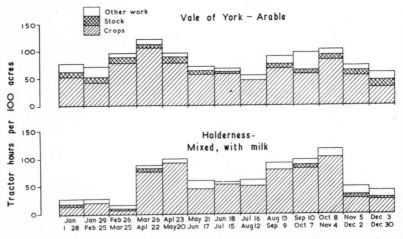

Fig. 7. Seasonal distribution of tractor work on two types of farm in Yorkshire.

Source: Leeds Univ. Dept. Agric. Economics.

flints, where use of the large rear-wheel-drive tractor reached high peaks in September and October, but fell off rapidly later in the year.

As Fig. 7 shows, seasonal irregularities may also be marked on some types of mixed farms. On arable farms, peaks in demand are natural to our climate and systems of farming, and while they may be minimized by suitable cropping policies, and will vary from year to year with the farming and climatic conditions, they will always be with us to a greater or lesser extent. Compared with some countries, such as Canada, these peaks and depressions are not severe, but it is not unusual to find a range from a peak of 160 hours per month in April down to less than 40 hours per month in February and June.

When the various types of tractors are studied, it is characteristic to find the tracklayers with more pronounced peaks and a sharper mid-summer depression than the wheeled tractors, while often doing more field work during the difficult winter months.

Type of Farming and Tractor Utilization *

Type of Work	Vale of York Arable	Holderness Mixed with Milk
	%	· %
Ploughing	18	34
Other field cultivations	29	27
Hay and silage making	6	9
Corn harvesting	10	13
Root harvesting	14	0
Stock work	6	3
Carting F.Y.M.	10	7
Other work	7	7
Total	100	100

* *Source:* Leeds Univ. Dept. Agric. Economics.

The type of work done by small–medium general-purpose tractors varies widely from farm to farm, and appreciably between groups of different farming type, as shown by average figures for two areas of Yorkshire, in each of which the tractor work is primarily concerned with crop production.

Annual Use and Cost of Operation

It is impossible to state with any accuracy the annual or hourly cost of running any particular kind of tractor, without first defining the work the tractor does. Its annual use may lie somewhere between nothing and 2500 hours a year, and one tractor may spend most of its time on light work, while another is generally given tough jobs. Costs will clearly depend on the amount and kind of work the tractor does, on the treatment it receives, and on its first cost, its age and the method of charging depreciation.

Average annual figures for different kinds of tractors, arranged according to annual usage are given in Table 12.

Most of the available surveys of tractor operating costs provide information which is obsolete, and changes in fuel prices may necessitate adjustment of the figures given here.

With 'gross margin' calculations it is sometimes desired to assess tractor fuel costs. These will depend on the type of work done. Typical figures of hourly consumptions of small–medium

Table 12. Typical Tractor Costs According to Type and Annual Use

Type 1. Wheeled small/medium – 45 h.p. Price £1600

	Hours worked annually				
	500	750	1000	1500	2000
Assumed depreciation life, Yr.	12	12	10	7	6
Annual fixed costs, £					
Depreciation	133	133	160	228	267
Interest, licence, insce.	75	75	75	75	75
Total annual fixed costs, £	208	208	235	303	342
Annual running costs, £					
Fuel,* ¾ gal (0·34 l)/h, oil & grease	94	141	187	282	374
Repairs & maintenance	80	108	128	168	208
Total annual cost, £	382	457	550	753	924
Hourly cost, £	0·76	0·61	0·55	0·50	0·46

* Fuel at 22 p/gal.

Table 12—*continued*

Type 2. Wheeled large/medium – 65 h.p. Price £2000

	Hours worked annually				
	500	750	1000	1500	2000
Assumed depreciation life, Yr.	12	12	10	7	6
Annual fixed costs, £					
Depreciation	167	167	200	286	333
Interest, licence, insce.	95	95	95	95	95
Total annual fixed costs, £	262	262	295	381	428
Annual running costs, £					
Fuel, 1 gal (4·5 l)/h, oil & grease	121	182	242	363	484
Repairs & maintenance	100	135	160	210	260
Total annual cost £	483	579	697	954	1172
Hourly cost, £	0·96	0·76	0·70	0·64	0·59

Type 3. Large four-wheel-drive – 95 h.p. Price £4000

	Hours worked annually				
	500	750	1000	1500	2000
Assumed depreciation life, Yr.	12	12	10	7	6
Annual fixed costs, £					
Depreciation	333	333	400	571	666
Interest, licence, insce.	185	185	185	185	185
Total annual fixed costs, £	518	518	585	756	851
Annual running costs, £					
Fuel, 2 gal (9 l)/h, oil & grease	231	347	462	693	924
Repairs & maintenance	200	272	320	420	520
Total annual cost £	949	1137	1367	1869	2295
Hourly cost, £	1·90	1·52	1·37	1·25	1·15

Type 4. Medium tracklayer with linkage – 80 h.p. Price £5500

	Hours worked annually				
	500	750	1000	1500	2000
Assumed depreciation life, Yr.	12	12	10	7	6
Annual fixed costs, £					
Depreciation	458	458	550	786	917
Interest, licence, insce.	250	250	250	250	250
Total annual fixed costs, £	708	708	800	1036	1167
Annual running costs, £					
Fuel, 2 gal (·9 l)/h, oil & grease	236	355	472	708	944
Repairs & maintenance	275	375	441	577	716
Total annual cost, £	1219	1438	1713	2321	2827
Hourly cost, £	2·44	1·92	1·71	1·55	1·41

tractors are given below. Where consumptions below about 0·4 gal per hour (1·8 l/h) are recorded, it indicates that a considerable amount of the tractor's time is spent idling.

Operation	Fuel Consumption per hour	
	gallons	litres
Rotary cultivating	1·0	4·5
Ploughing, cultivating, disc harrowing	0·8	3·6
Combining, potato lifting	0·7	3·2
Harrowing, rolling, grubbing, ridging, drilling, binding, potato planting and belt work	0·6	2·7
Hoeing, covering potatoes, light harrowing, distributing fertilizer and manure, mowing hay and silage, operating pick-up baler	0·5	2·3
Spraying	0·4	1·8

Source: East of Scotland Agricultural College.

Large tractors, which are used mainly for heavy work, tend on average to be used at a higher percentage of their potential power than smaller tractors. Thus, typical overall fuel consumptions are $\frac{1}{2}$ gal (2·2 l)/h for a 40 h.p. tractor, 1 gal (4·5 l)/h for a 65 h.p., 2 gal (9 l)/h for a 90 h.p. and $2\frac{1}{2}$ gal (11·2 l)/h for 100 h.p. Maximum consumptions are about four times as high for the small tractors and about twice as high for the large.

Operating Speeds and Type of Transmission

Forward speed naturally has an important bearing on the overall rate of work, so it might be expected that the highest speed practicable is normally chosen. This is not necessarily so, for a variety of reasons, chief of which are quality of work and operator comfort. Provided that these are satisfactory, there is usually little to be gained by running tractors at low speed if they can go faster. A good range of gears enables a conscientious driver to get the most out of the tractor. It is a mistake to suppose that the tractor will last a great deal longer if it is not fully loaded. On the other hand, harm can soon be done by persistent over-loading.

The following is a rough guide to the speeds commonly needed for different kinds of work with medium-powered wheeled tractors.

(1) *Very Low Speeds* $\frac{1}{2}$ to 2 m.p.h. (0·8 to 3·2 km/h)
  Transplanting
  Potato planting   Use of complete potato harvester
                                 Mole draining

(2) *Low Speeds* 2 to 4 m.p.h. (3·2 to 6·4 km/h)
  Ploughing   Close hoeing   Ridging
  Cultivating   Digging potatoes   Use of sugar-beet harvester
  Rolling   Spaced drilling   Use of forage harvester
  Combine harvesting   Use of pick-up baler

(3) *Medium Speed* 4 to 6 m.p.h. (6·4 to 9·6 km/h)
  Light cultivations   Mowing   Swath turning and tedding
  Weed spraying
  Drilling   Fertilizer distributing   Field haulage

(4) *Fairly High Speeds* 6 to 8 m.p.h. (9·6 to 12·9 km/h)
  Field haulage in good conditions

(5) *High Speeds* 8 to 20 m.p.h. (12·9 to 32 km/h)
  Road haulage and running light.

Selection of the most suitable gear and governor setting can have an important effect on fuel consumption, especially on light work. A high gear and low engine speed will usually give improved fuel economy over a low gear and 'normal' engine speed.

One of the major problems in choosing a tractor is assessing the need for and value of the differing types of transmissions now offered. It is clearly advantageous if the tractor can provide a range of speeds from about $\frac{1}{2}$ to 15 m.p.h. (0·8 to 24 km/h) or more, with ability within the working ranges to vary speed without appreciably altering power output from the engine. In practice, this is usually achieved by a combination of several gears and use of a flexible governor, which functions over a wide range of engine speeds. A feature of most engines which has to be accepted is that power output will inevitably fall off at low engine speeds. The more gears that can be provided, and the easier it is to change gear, the more likely it is, provided the gear ratios are well chosen by the designer, that the tractor will be able to give a good performance at any chosen speed.

Tractor transmissions have been developed, by steady progress, from three-speed gear boxes, first to four, five or six gears, and more recently to ten or twelve. It might be thought that there is an element of 'keeping up with the Jones'' in this; and certainly progress would not have been as rapid but for the keen competition between manufacturers. The fact is, however, that for the farmer who has a wide range of field work to do, and expects his tractor to be a 'maid of all work', the 10–12 speed gear box is the fulfilment of a genuine need, and not of a whim.

Provision of a 'change-on-the-move' transmission can be well worth its extra cost provided that it is made use of. It should have the effect not only of providing the physical means to increase tractor output, but also of giving a psychological stimulus to the operator. It has been estimated that in heavy work, power-shift can result in an increase in output of up to 20 per cent. Where such an increase can be achieved, the additional investment can hardly fail to be justified, provided the right size of tractor is chosen, and the driver makes good use of the facilities provided.

Power-assisted mechanical transmissions have gone far towards achieving the objectives of 'step-less' hydrostatic transmission. Moreover, they are reasonably efficient over a very wide range of duty, whereas simple hydrostatic transmissions tend to be efficient only over a more limited range of duty.

Optional Fittings

With modern British mass-produced tractors, the basic hydraulic system is so much a built-in part of the basic tractor that there is no question of whether a farmer should buy it. It now normally includes a depth control for mounted implements, most manufacturers having come round to the view that built-in hydraulic draught control is a desirable feature.

There are, however, differences between manufacturers in what is offered as standard equipment, and in the control systems provided. This is an involved subject, which is more fully discussed elsewhere.[13]

Optional fittings include such devices as the equipment provided by one manufacturer to enable the 'pressure control' system to assist in transferring weight from trailed implements to the rear wheels of the tractor. As a general rule, farmers should assume that where a manufacturer only offers equipment as an optional fitting, it is considered that there are some farmers who will not obtain a worth while advantage by having it fitted. Each optional item should therefore be considered separately, since there may be a good case for economizing where the fitting will very seldom be used. A 'constant running', 'live' or 'independent' p.t.o. is now almost an essential on farms where the tractor operates a range of power driven machines, including hydraulically operated equipment such as front loaders. Independent drive is useful on machines such as mowers, balers and forage harvesters, since it is often useful to be able to check forward motion without stopping the drive to the machine. There are, however, farms which seldom use a tractor for any of these purposes; so some manufacturers only provide a dual clutch or other means of providing a 'live' drive as an operational extra.

Tractor Types

Many complex technical considerations are involved in buying a tractor for the range of needs on a particular farm. These cannot be discussed in detail, but mention must be made of points that need to be studied.

Medium-powered all-purpose tractors are usually the best 'value for money', owing to their mass production. Very large, wheeled tractors may be economic on large light-land farms where they can be effectively utilized for all cultivations, drilling and harvesting; but they tend to be more expensive in relation to their effective power. (See Table 11.) At the other end of the scale, very small tractors also tend to be expensive in relation to output, and second-hand medium-powered tractors are a better proposition except in the few cases where the small machine has important technical advantages. The driver's time is as important on small farms as on large ones, and when it is possible to buy a reasonably good medium-powered tractor for under £500, there is little to be said in favour of a very small new one costing over twice as much.

As regards *special-purpose* tractors, extreme row-crop types or self-propelled tool carriers may justify their use for specialized work on large farms, but the rear-engined tractor has not sufficient advantages over the conventional type to become widely used for general purposes in its present forms. So far, nobody has been able to invent implements and machines which are as easily attached as those which go on the 3-point linkage; and some items, such as drills and hoes, take a considerable time to put on to the rear-engined type. Skeleton or cage wheels are useful extra equipment for certain types of work, and there is often advantage in using them now that wheels are fairly easily changed. The cost of such accessories is small in relation to the potential benefits, and the saving of wear on rubber tyres.

Tracklayers of all kinds are comparatively expensive to operate as well as to buy, but a medium-powered machine can pay for itself on heavy land, due to ability to operate in adverse conditions, and to get the work done in season. A tracklayer with a diesel engine of 75 h.p. costs about twice as much to buy as a wheeled tractor with a similar engine. Its operating cost, on a basis of about 1000 hours annual usage, is also almost double that of the wheeled tractor, mainly owing to the higher depreciation and occasional incidence of expensive track repairs. (See Table 12.) It is, therefore, important to avoid using the tracklayer in conditions where a wheeled tractor will work equally well; for where the conditions for wheel grip are satisfactory, the wheeled tractor can do as much work as the tracklayer.

Heavy tracklayers are very expensive both to buy and to operate, and are chiefly suitable for use by contractors. The guiding principle should be to use the cheapest equipment that is reasonably effective. There are marked economic advantages in using wheeled tractors rather than track machines wherever they are effective, and much turns on the ability of the tractor wheels to transmit power without excessive slip. *Four-wheel-drive* is attractive from a technical aspect and considerable advances have been made in applying this principle to large–medium and high-powered units which are used principally for heavy cultivations.

High-powered Wheeled Tractors

Mention has already been made of the fact that high-powered tractors tend to be expensive in terms of capital cost per unit of power. Before considering other economic aspects of their use, it is useful to review typical performances in practical operating conditions.

Different types of large tractors differ in their adaptability to a range of duties. In general, rear-wheel-drive tractors of conventional design tend to be most adaptable to a wide range of jobs: 4-wheel-drive machines with moderately small front wheels are reasonably adaptable; and tractors with equal-sized wheels are used mainly for soil cultivation, drilling and forage harvesting. It is therefore necessary to consider for each individual farm whether the large tractor is expected to do such jobs as fertilizer distributing, spraying, mowing or the many other light jobs involved in making hay or silage. On some mixed farms there may be a real advantage in being able to put a high-powered tractor on to a job such as flail mowing, or using a full-chop forage harvester; for this is heavy work which needs to be done at high speed when the crop is ready. There is therefore an advantage on such farms if the high-powered tractor is of conventional type, with a good live p.t.o. On most of the lighter work, such as spraying, fertilizer distributing, tedding, etc., the high-powered tractor has no technical advantages over much smaller machines; and there are often good physical reasons, such as avoidance of soil compaction, why use of a lighter and less powerful tractor is preferable.

The main reason for buying a high-powered tractor is normally to achieve faster work on heavy cultivations. Most weight should therefore be given to efficiency at such work. Information comes mainly from user experience and to a lesser extent from official tests. Such evidence as is available shows a high efficiency for 4-wheel-drive tractors with four large drive wheels, especially where the work is on heavy land. Efficiency tends to be lowest for tractors with only rear-wheel-drive. Thus, there is a tendency for efficiency at heavy work to be in the reverse order to degree of adaptability for a wide range of lighter work. When considering choice of equipment, due account should be taken of the influence of soil characteristics on performance. Thus, on wet, heavy soils, adhesion and soil compaction can present serious problems. On such soils, where draught is likely to be double that of light land, the superior pulling ability of the tractors with four large drive wheels is likely to be most beneficial. On some lighter soils adhesion and the performance of rear-wheel-drive tractors with very large tyres can be entirely satisfactory. Ability to exploit high tractor power naturally depends on providing suitable implements, and on going fast when conditions allow. A very large tractor (over 100 h.p.) should give a performance in line with the following figures in average conditions:

Operation	Working depth in	cm	Speed m.p.h.	km/h	Output (overall) acres/h	ha/h
Plough heavy land 5-furrow, 14 in	8	20	3	4·8	1·7	0·7
Plough light land 6 furrow, 14 in	7	18	3½	5·6	2·3	0·9
Deep cultivate heavy land (10 ft wide 9-tine)	10	25	3	4·8	3·0	1·2
Disc harrow (15 ft wide)	3	7½	4	6·4	5·8	2·3
Spring-tine cultivator/harrow (20 ft wide)	4	10	4	6·4	7·7	3·1
Rolling (30 ft set)			5	8	14	5·7

In particularly favourable conditions with wider implements, higher rates of work may be possible; but in practice it is difficult to achieve higher rates on typical British farms, owing to such factors as limited field sizes. If full use is to be made of high-powered tractors, field sizes must not be too small, and it is worth while to do some enlarging where practicable. A size of about 50 acres (20 ha) should be the aim. At this size it is practicable to complete an operation such as drilling of the field in a normal day.

Economics of High-Powered Tractors

When attempting to assess the economic effects of buying and using a high-powered tractor, there is no satisfactory alternative to studying the question very broadly, on a 'whole-farm, whole-year' basis, since the effects involved go far wider than spending a certain amount of money in order to achieve a higher rate of work. The main factors involved are:

> extra cost of owning and operating the high-powered tractor, including the extra cost of new implements;
> saving of labour on a whole-year basis;
> increased output due to improved 'timeliness'.

Capital cost of the high-powered tractor will be of the order of £4000 or more. In addition, it will probably be necessary to spend an additional £2500 on a suitable large plough, heavy cultivator and spring-tined cultivator-harrow. As a general rule, it will be difficult to justify such expenditure on a farm with less than about 400 acres (160 ha) of arable land. Table 12 (type 3) gives an indication of the operating cost, which tends to be excessive at an annual use of less than about 750 hours. The main economic effect of introduction of such a tractor turns on whether the farm is able to reduce the regular labour force by one man. This, in turn, depends on whether the size of the labour force is determined by needs at times of the year when use of the high-powered tractor is effective, e.g. in autumn or spring, when it should be working almost continuously on ploughing and/or seed-bed preparation.

In circumstances where it is possible to replace two large–medium-powered tractors by one very large one, it can be roughly reckoned (Table 12) that the costs of running the very large one will be approximately the same as that of running two large–medium-sized ones. Similarly, the cost of operating the equipment pulled by the high-powered tractor is likely to be roughly the same as that of the two sets of equipment for the smaller tractors. Differences in the 'overheads' will be insignificant, since capital cost of the two is equal to that of the high-powered machine. Where use of the large tractor makes it possible to run the farm with one man less, the reduction in cost is considerable, e.g. £1500 per year. Surveys show, however, that this is seldom achieved immediately. It is most likely to be possible on very large arable farms where a high proportion of cereals is grown. It is improbable on farms where there are high labour peaks in summer, at a time when the high-powered tractor is likely to spend much of its time either idle or on light work.

The value of improved timeliness can only be assessed on individual farms. Some reference has already been made in Chapter 2 to the effects of sowing date on the yield of winter wheat. Similar effects are produced in many other farming operations, and there can be no doubt about the general importance of getting work done in good time.

Efficient Tractor Operation

Tractors have become so much a part of farm work of all kinds that their efficient operation is of the utmost importance. In addition to the obvious need for regular maintenance, it is increasingly necessary for farmers to give some thought to possibilities of reducing working time by cutting out unnecessary journeys with very small loads, and by working at the highest practicable speed. Generally speaking, few British farmers exploit the possibilities of combining jobs by the use of multiple hitches; and development of the use of mounted implements tends to reduce opportunities for drawing a train of implements. Nevertheless, there are sometimes occasions when the harrows could follow the drill or the roll, and when side-rake or tedder and mower could be used at the same time. Ground conditions often limit forward speed, but here again, the necessary rolling in spring will allow the mower or forage harvester to be used at an appreciably higher speed. As a general rule, more use could now be made of the wide range of gears and engine speeds provided on modern tractors. There is a tendency for tractors to be unnecessarily under-loaded on cultivations by working in a low gear; and while it is not suggested that speeds should be raised to the limit of safety, a careful study of machines at work can sometimes lead to an advantageous stepping up from one gear to the next.

Power for Stationary Work

The extension of electricity supplies to farms has proceeded rapidly since 1945 and only the most inaccessible now lack a public supply. Electricity is almost essential for efficient mechanization of work about the buildings, and it has a wide range of uses. Most of the stationary power jobs about farm buildings can be more economically done by electric motors than by any form of internal combustion engine. The advantages of electrical power include low capital cost and a very low rate of depreciation; cleanliness and convenience; ease of control, either by hand switch or by various kinds of automatic devices; constant availability, and a minimum of maintenance and repair work. The advantages of electrical drive are perhaps most appreciated when it becomes necessary to provide power for a fairly large number of inter-related barn machines such as elevators, a grinder, a crusher and a mixing machine. Electrical drive to individual machines makes it possible to place the machines just where they are wanted, and not where a line of shafting requires them to be.

Comparative Costs of Diesel and Electric Equipment

Though most farms now have some kind of public electricity supply, not all have the kind of supply which enables them to operate large motors. The existing power lines, often of single-phase type, were designed primarily for bringing a limited supply to a large number of consumers, and may need very expensive reinforcing before powerful crop-drying units can be installed. Occasionally, however, when a reasonably adequate supply is available, it may be necessary to compare the capital and operating costs of electrical and diesel engine drive units. In such circumstances due weight should be given to the very great advantages of electric motor drive already mentioned. When making cost comparisons, it should also be borne in mind that a fan duty requiring use of a 30 h.p. electric motor will call for a diesel engine of about 40 h.p., since experience shows that continuous loads such as driving a crop-drying fan or pumping tend to impose excessive strain on the engine if it is loaded much above about 80 per cent of its maximum capacity. Electric motors, on the contrary, can operate indefinitely at full load. Bearing in mind these provisos, operating costs of comparable diesel and electric crop drying units, used for 1000 hours annually may be reckoned as in Table 13. In this example it is assumed that the waste heat from the diesel engine is beneficial and is efficiently utilized; and that to secure a comparable performance, the electric unit needs to use a heater when drying conditions are adverse.

The calculation shows that even in a season when use is high (1000 hours) and about 10 kW of heat is used on average, the

Table 13. Comparative Annual Costs of Operating Diesel and Electric Crop Drying Units. Based on 1000 Hours Annual Use

	Diesel Unit	Electric Unit
Specification	40 h.p. engine with fan, arranged to utilize waste heat	30 h.p. motor; 60 kW heater; switch gear
Capital cost £	1200	1000
Depreciation: Basis	20% p.a. on complete unit	10% p.a. on complete unit
Annual Cost £	240	100
Interest on capital @ 8% p.a. on average investment. £	48	40
Fuel and Power: Basis	2 gal (9 l)/h for 1000 h 22p per gal	Average 36 kW per hour * for 1000 h @ 1·2p per unit
Cost £	440	432
Lubricating oil, Repairs & maintenance. £	50	10
Total Annual Cost £	778	582

* Includes average of 10 kW for heater.

electric unit has a lower total cost, mainly owing to lower depreciation.

In estimating the capital cost of an electrical installation, account must be taken of any extra supply charges that may be involved, since the farmer may have to contribute a substantial sum towards the provision of an additional supply. This can sometimes tip the scales in favour of the diesel engine, which can also have the considerable advantage of mobility if it is mounted on a suitable chassis. Electricity has important advantages: no refuelling; very little maintenance; easy automation; and less noise.

CHAPTER SIX

Crop Production Mechanization

Cultivation Implements. 'Minimum' Cultivations. Rotary Cultivation. Manure Handling. Farmyard Manure Loading and Spreading. Slurry Disposal. Fertilizer Distribution and Drilling. Inter-row Hoeing. Down-the-row Thinners. Potato Planting. Potato Chitting. Use of Transplanting Machines. Field Crop Spraying. Irrigation.

Cultivation Implements

The figures for tractor utilization already quoted (Chapter 5) give some idea of the importance of basic cultivations on arable farms. Moreover, there are many farms where the peaks of autumn and/or spring work are so high that tractors and men are more than fully occupied. This means that there are real advantages in finding ways of increasing output per man on cultivations.

Studies of average rates of work show that performance is often far from what should be possible. There are many reasons for this, including lack of understanding of the tractor's capabilities by the driver; lack of incentive to work fast; and sometimes lack of such necessities as a good tractor seat to permit fast work to be done in safety and comfort. One of the main obstacles to efficiency is the difficulty of matching the width of implement to the tractor's power and speed. Often, implement width has to be a compromise between providing a wide enough implement to make use of the tractor power at a reasonable forward speed, and avoiding spending too much money on the implement, or running the risk of being unable to pull it due to wheel slip in adverse conditions. The modern trend must generally be to use as high a forward speed as practicable, in order to keep down implement cost and avoid tractor wheel adhesion difficulties.

For ploughing, $2\frac{1}{2}$ m.p.h. (4 km/h) was formerly regarded as a normal speed. This is usually too low for modern tractors. 3–4 m.p.h. (5–6·5 km/h) may now be regarded as normal, and speeds of 4–5 m.p.h. (6·5–8 km/h) can be advocated where the going is good and the tractor has the necessary power. Still higher speeds may be practicable in some circumstances, in light soil and with suitable high-speed plough bodies. Plough draught tends to increase as speed is raised; but this disadvantage may be offset where a seed-bed is to be prepared soon afterwards, since faster work tends to result in more pulverization of the furrow slice.

When ploughing in good conditions a field efficiency of about 82 per cent should be attainable. Soils vary widely in their resistance to working by cultivation implements. For example, while average resistance to ploughing of arable soils in good condition is about 8 lb per sq in (about 55 kN/m²), it may range from half this figure in very light soils to twice it in heavy clays. Resistance per unit area tends to increase a little with increase of ploughing depth. A 65 h.p. tractor should be able to handle a 3-furrow plough on deep work in most soil conditions, and should be able to handle 5–6 furrows at shallow depth in light soils. The type of plough most suitable for the shallow work is likely to be a semi-mounted one, while the 3-furrow plough for deep work could well be a fully mounted reversible one.

For very large tractors, provision of suitable ploughs is a difficult problem. A 130 h.p. tractor should be capable of pulling 8–10 furrows at shallow depth. Very large semi-mounted ploughs with up to 8 furrows have been developed, but these are, naturally, expensive.

Shallow-working tined implements such as a spring-tined cultivator/harrow have a draught of only about 100 lb per ft (1·46 kN/m) width at 3 in (7·5 cm) depth; so a 20 ft (6 m) wide implement pulled at 5 m.p.h. (8 km/h) only needs about 13 drawbar h.p. to operate it. This means that though a large tractor will give a good performance with the implement, a much smaller tractor would be adequate for the job.

Depth of Cultivation

The optimum depth of cultivation for the various crops is a matter which must usually be decided mainly by type of soil, climate, topography, drainage, cropping policy and a whole host of factors which are extremely difficult to assess from the standpoint of economics. Where the subsoil consists of solid limestone, large lumps of chalk or inert rock, the depth of cultivation is often very restricted; but in other conditions, where drainage is satisfactory and there is a good depth of ploughable soil, some farmers favour very deep cultivations while others prefer a moderate or shallow depth of work. In Britain, it is safe to say that a majority of the best farmers plough deep occasionally, and experiments show some slight advantages in yields for deep work provided that additional fertilizer is added to spread readily available nutrients throughout the larger volume of soil worked.

The results of a series of experiments in the early 1950s in which very deep ploughing and subsoiling were employed have been summarized as follows.[15]

Crop	Proportion of Experiments in which Deep Cultivations produced	
	Increased Yields	Decreased Yields
Sugar beet	51%	26%
Potatoes	37%	22%
Cereals	44%	18%

The comparison was usually between 14-inch (36 cm) and 8-inch (20 cm) ploughing, and after the experimental cultivations the first crop was roots, and this was followed by wheat.

When the results were analysed according to soil type, clays and loams were found most responsive to deep cultivation, and half of such fields gave increased yields as a result of either deep ploughing or subsoiling. On sands, on the other hand, deep ploughing decreased yield on 40 per cent of the treated fields and seldom showed much benefit. In light loams, subsoiling was beneficial, and increased yields on about half of the fields.

On heavy land, the deeply ploughed plots tended to dry out earlier in spring, and in some circumstances work could begin a fortnight earlier on these plots.

One advantage of deep work is in respect of weed control; for really deep ploughing will greatly reduce the annual weed population, and will often entirely eliminate troubles from perennial weeds.

More recent work at the Ministry's Experimental Husbandry Farms has shown that there is no worthwhile gain where there is no particular reason for deep work. The general trend of results is the same for a wide range of soil types; it is that deep ploughing (e.g. 15 in (38 cm) on deep heavy silts at Terrington; 12 in (30 cm) on Hampshire chalk) has no advantage over ploughing at moderate depth (9 in (23 cm) at Terrington; 8 in (20 cm) on chalk); but ploughing at moderate depth is generally preferable to very shallow work (e.g. 6 in (15 cm) deep for potatoes, 4 in (10 cm) for other crops).

The extra cost of deep cultivations may be considerable. If 6–8 inch (15–20 cm) ploughing costs £2 an acre (£5/ha), 12–14 inch (30–36 cm) work will almost certainly cost at least double, and may, moreover, necessitate the use of very expensive equipment, such as tracklaying tractors and very heavy ploughs. On the other hand, it is often possible to plough 8–10 in (20–25 cm) deep with wheeled tractors and equipment that is not expensive, and many farmers favour such work for root crops. As indicated above, deep ploughing often results in less trouble from weeds, and the extra cost of the initial cultivations may be balanced by less expense on after-cultivations.

In the dry-land wheat-farming areas of the Great Plains of U.S.A. and the Canadian prairies, where the rotation is often (1) wheat, (2) fallow or (1) wheat, (2) wheat, (3) fallow, use of the mould-board plough has been discontinued, and the depth of cultivation now generally practised is as little as 2–3 in (5–7·5 cm). The common cultivation implement is the 'one-way discer', and throughout the low-rainfall area 'stubble-mulch' farming is practised. The stubbles are seldom touched in autumn, and after the frost and snow has cleared in spring the wide one-way discer with seed-box attached gets busy. Use of this implement permits very early drilling, and the stubble that is left sticking out of the land helps to protect the soil from wind erosion. The fallows are worked with the same implement just sufficiently to check weed growth and so reduce loss of moisture by transpiration. This example is quoted merely to show that there can never be a clear-cut answer to the question whether deep cultivation pays. On many British farms, it is a great advantage to be able to plough and cultivate deeply *when necessary*. But even on such farms, it may often be economically advantageous to work the soil only lightly for certain crops.

'Minimum' Cultivations

Analysis of the results of experiments on cultivations shows that one of the basic reasons for most cultivation practices is effective weed control. Developments in chemical weed control have now advanced to a point where it is possible, though not necessarily most economic, to control most weeds by spraying, rather than by cultivation. The question therefore arises whether cultivations as such have any economic value beyond weed control. Long-term experiments on continuous cereal growing with 'restricted' cultivation and also with the minimum necessary to get seed covered have therefore been carried out, to discover the effects on both crop yields and costs. Results indicate that it is unlikely that extreme lack of cultivation will be found economic when continued without interruption over a long period. Though it is difficult to prove the value of cultivation by experiment, it will probably be found that the problems created by long-continued lack of cultivation can be quite serious on many British soils, and that by comparison, cheap and simple operations such as shallow ploughing and cultivating are economically advantageous. For the present, prolonged absence of cultivation in any case seems doomed to failure, if only because of the difficulty of controlling grassy weeds economically by means of herbicides. There are, however, circumstances where the use of a chemical such as paraquat in preference to cultivating, is clearly advantageous; e.g. where there is a good surface tilth on heavy land, and cultivation is certain to spoil it and make seed-bed preparation difficult.

The comparative costs of conventional cultivation and the

technique of using a herbicide such as paraquat, followed by direct drilling, depend on the type of equipment used. Contract prices are not a suitable basis for assessing cultivation costs, since most farmers have their own equipment and can do cultivations quite cheaply. Ploughing and two strokes with a spring-tine cultivator-harrow may be regarded as normal cultivation. The cost of these operations should not exceed about £5 per acre (£12·50/ha).

By comparison, the spray chemical is likely to cost about £4 per acre (£10/ha), and applying it may be reckoned at about £0·50 per acre (£1·25/ha). Cultivations are therefore likely to be almost as cheap on most farms, even if drilling after the chemical does not cost any more than drilling after conventional cultivation.

Nevertheless, there is reliable evidence that cereal crops do not require very much tilth, and that time and money can often be saved by avoiding unnecessary tillage. Avoidance of ploughing, in particular, can often appreciably reduce cultivation costs and at the same time lead to improved timeliness of sowing. If in these circumstances a light application of a herbicide such as paraquat is needed to eliminate the danger from seedling weeds which have been missed by the reduced cultivation system, this is a small price to set against the advantages. Independent trials on experimental husbandry farms show that there are economic advantages in well-chosen and well-executed 'reduced cultivation' systems for cereal growing.

For root crops such as potatoes, thorough though not over-deep cultivation before planting is essential: but inter-row after-cultivations are only of value for weed control and avoidance of greening. Comparisons between after-cultivation and use of herbicides at Terrington E.H.F. showed substantial yield increases from use of herbicides and avoidance of cultivation.

Power-Driven Cultivators and Harrows

The advantages of the rotary cultivator in some circumstances can hardly be questioned. It will make an excellent job of corn stubbles after the combine harvester, where ploughs, cultivators or even discs would be troubled by straw; it will make a seed-bed after kale or brussels sprouts without preliminary disposal of the stalks; it can do an excellent job if used before the plough in preparing land for potatoes after a long ley; and on more normal duties such as seed-bed preparation after ploughing, a rotary cultivator can in favourable conditions work down a tilth without loss of moisture in a manner that would be impossible with cultivators and harrows. In adverse conditions rotary cultivators can sometimes 'force' a tilth suitable for drilling, where even disc harrows have little effect. Some former limitations and disadvantages have been removed by the provision of fixed tines which can be set to break up any pan that tends to be formed in some soils by the action of the common type of L-shaped rotor blade: and for seed-bed preparation 'spike-tooth' rotors can be fitted.

Rotary cultivators now have to compete with a variety of power-driven harrows for seed-bed preparation. As a general rule, these power harrows, whether of the reciprocating or rotary type, are designed to work after a primary cultivation which is usually either ploughing or comparable work produced by deep working with a heavy cultivator or 'chisel plough'. The great advantage of power-driven implements is that it is usually possible to produce the desired effect in one passage across the field. Furthermore, the unwanted compaction produced by tractor wheels on a partly worked seed-bed can be avoided.

Since power-driven cultivators and harrows tend to be more expensive than the simpler implements they replace, they should be used judiciously. There are occasions when the tilth-making characteristics of the power-driven machine are not needed. Where it is really needed it is invaluable, but where a stroke with a simple implement, e.g. a spring-tined cultivator, will do a satisfactory job, this will usually be cheaper.

Manure Handling

Until livestock production was intensified, there was little argument about the desirability of making farmyard manure, and returning it to the land to assist in crop production. In some mainly arable areas there used to be a tendency to judge the standard of farming by the amount of farmyard manure applied. Today, however, values have changed. Labour is relatively more expensive, and many experiments have shown that while organic manures can have marked beneficial effects on the structure of certain types of soils, most soils can remain productive without such manures, provided that adequate amounts of the right chemical fertilizers are applied. When once the idea that farmyard manure is not essential as a source of humus is accepted, it becomes necessary to compare the cost and labour involved in handling it with that of avoiding this work, or finding ways of minimizing it.

Where straw is expensive, there is clearly little to be said for making conventional farmyard manure; and methods of handling the dung and urine as a slurry need to be examined. This does not mean that handling manure as a slurry is always the best solution. In fact, where there is plenty of home-grown straw available, there are many good reasons for using it for bedding. Advantages include avoidance of effluent problems, and the fact that farmyard manure can be easily stored, either where made or in the field. It can then be applied to the land when there is time to spare, or when the land is ready to receive it. Most slurry handling systems, on the other hand, impose a rigid disposal time-table which can sometimes be extremely inconvenient.

It can be reckoned that when adult cattle are kept in a straw yard for about 5 months they will need about $\frac{3}{4}$ ton of straw per head, and each ton of straw will result in about 4–5 tons of manure to be handled.

Farmyard Manure Loading and Spreading

Handling farmyard manure is a good example of the rapid advance of mechanization. Until about 1945, both mechanical loaders and spreaders were regarded as something of a luxury; and in the 1959 edition of this book it was considered worth while to devote space to comparisons of hand work and mechanical methods of loading and spreading. Today, with moderately priced equipment available to increase a man's output about twenty-fold, while eliminating all the hard manual work, moving manure by hand fork should be regarded as a misuse of labour, and unnecessary even on small farms. It does not follow, of course, that every small farmer can justify having his own loader and spreader; but the possibilities of contract work, equipment hiring, and machinery syndicates should be considered.

Though there are other methods, such as use of a dozer blade on a tractor, the equipment now most widely used for handling farmyard manure is a hydraulic loader and a trailer type self-emptying spreader. There are, however, many variations in types and capacities, and there are some circumstances where a hydraulic grab or rear loader, rather than a front loader, is chosen. Advantages of the more expensive hydraulic grab include less wear on the tractor clutch, gearbox and tyres, and less cutting up of the ground surface.

Where a yard is free from obstructions and has a firm bottom, output from a modern front loader on a powerful tractor, fitted with power-operated change-on-the-move transmission, can be up to 1 ton per minute with a skilled operator. This means that output depends largely on having sufficient transport available, and that most handling systems cannot keep the loader fully occupied. Where it is practicable to provide several large transport vehicles, an overall loading rate of up to 30 tons per hour is attainable.

Spreading speeds usually range from $2\frac{1}{2}$ to 6 m.p.h. (4 to 10 km/h), depending partly on the dressing applied. The largest machines can spread up to 3 tons per minute (spot working rate), but more common rates are $\frac{3}{4}$ to 2 tons per minute. Spreading width depends on the size and type of spreader. It tends to be fairly narrow (e.g. 7–8 ft (2·1–2·4 m)) with most spreaders which have the beater mechanism at the rear of a conventional type of trailer body. With machines having a flail type mechanism which spreads to the side of a cylindrical body, effective spreading width is usually about 8 ft (2·5 m). Large p.t.o.-driven machines with multiple beater mechanisms may have a wider bout width, e.g. up to 15 ft (4·5 m). The multiple beater type may therefore be preferable where it is desired to limit the number of wheel tracks on soft land.

Before considering choice of loading and spreading equipment, emphasis should be laid on the importance of doing whatever is practicable to make the buildings suit the handling methods. Often, a little time and cost spent on altering yards to suit the use of a hydraulic loader is as important as choice of loader.

Choice of Manure Handling Equipment and System

On most farms the vital factor affecting choice of handling systems is the number of men available. Often there will be only one or two. One of the best systems for a small farm is a 'one-man' system with a land-wheel-drive spreader which is set down for loading, and picked up on the hydraulic pick-up hitch. (System 1, Table 14.) Ease of hitching makes it desirable to avoid the complication of p.t.o.-drive. Capacity of such spreaders is generally limited to about 2 tons, partly because of the extra draught involved in operating the spreading mechanism from the land wheels. Because of this limitation, some farmers prefer to use a second tractor to operate the loader. (System 2.) With the limitation of ground-wheel drive removed, a larger-capacity p.t.o.-drive spreader is a possibility.

Where two men are available, these should ideally have a large-capacity p.t.o.-drive spreader each, and there should be a third tractor equipped with a hydraulic loader (system 5); but there are other effective systems (3 and 4) of operating smaller spreaders. Where one of the spreading tractors operates the loader (system 3) the two spreaders work together. Where a third tractor is used to operate the loader, each man works independently, and working rate is considerably improved due to avoiding hitching and unhitching, and waiting time.

It is usually difficult to justify the type of system which involves a man operating a loader continuously, except on very short hauls, e.g. from a field heap, where a 3-man system with two large spreaders can achieve a high rate of work. System 6, with 3 large spreaders, involves heavy capital expenditure, and is suited only to large-scale operations or contract work, where it may be necessary to move very large amounts in a short time.

Table 14 indicates the advantages open to machinery syndicates, which can choose large equipment suited to small gangs, and achieve all the benefits of large-scale operation.

Table 14. Farmyard Manure Handling Systems. Target Rates at $\frac{1}{2}$ Mile Transport Distance

System No.	Men	Number of Tractors	Loaders	Spreaders	Load size tons	Loading	Minutes per Load Transport	Spreading	Hitching etc.	Waiting	Loads per Hour	Tons per Hour
1a	1	1	1	1	1¼	3	8	3	3	—	3½	4½
1b	1	1	1	1	2	4	8	3	3	—	3	6
2	1	2	1	1	4	8	8	4	—	—	3	12
3	2	2	1	2	2	4	8	3	3	4	5	10
4	2	3	1	2	2	4	8	3	—	—	8	16
5	2	3	1	2	4	8	8	4	—	—	6	24
6	4	4	1	3	3	6	8	4	—	1	9	27

Because of the importance of load size, and the high cost of spreaders compared with that of trailers, there is a good case for using large-capacity tipping trailers for transport over long distances, even if this necessitates double handling. Double handling is in any case often necessitated by crop husbandry considerations. Where the land is available to receive the manure and the surface is firm, there may be advantage in placing the trailer loads at the required intervals over the field.

Slurry Disposal

The amounts of slurry which are likely to be produced by various types of stock vary considerably with the housing conditions. Large dairy cows, for example, which on average produce about 10 gallons (45 l) of dung and urine daily, are usually kept in conditions where only a part of this has to be handled as slurry. On the other hand, in circumstances where all is handled by organic irrigation, this 10 gallons per day may be doubled by the addition of dairy washings, etc. In cowsheds, where about 7 lb (3 kg) per head per day of bedding straw is used to soak up the manure, there is likely to be only about 2 gallons (9 l) per head per day of liquid run-off. In a semi-covered yard, though a little more bedding straw is likely to be used, there will usually be about 3 gallons (14 l) per day of slurry to handle; and where there is any considerable area of concrete collecting yards, silage feeding areas, etc. this may often be increased by up to 50 per cent, especially in high rainfall areas. In fully-covered cubicle houses, where about 12 lb (5 kg) per head per month of sawdust is used, the full 10 gallons (45 l) daily of slurry will have to be handled, and this will rise to about 13 gallons (60 l) daily if washing water is added.

In the case of poultry, typical hybrid layers kept in battery cages produce about 1 ton of manure weekly per 1000 birds. This material has a moisture content of about 75 per cent and occupies about 30 cu ft (0·85 m³) or 200 gallons (900 l). Larger birds, dilution with spilt drinking water and less digestible foods may result in double this quantity. At 1 : 1 dilution, the mixture is sufficiently liquid to be dealt with by a good tanker.

Fattening pigs when meal-fed produce on average about $2\frac{1}{2}$ gallons (11 l) of dung and urine daily. Whey-fed pigs produce about 3 gallons (14 l) per head. Here, again, there may be dilution with spilt drinking water.

The main slurry 'problem' on most farms is finding a cheap and easy method of disposal, though investigations of the fertilizer value show that the handling cost can be repaid by savings on bought-in fertilizers where the slurry can be applied in suitable quantities at the right time. On most farms it is a considerable disadvantage to employ a system which necessitates spreading slurry almost every day. Generally, it is desirable to provide storage sufficient to hold at least a week's production, and an advantage to be able to hold much more if required. Unfortunately, the cost of steel or concrete storage tanks can be considerable, but such above-ground storage suits most farms, though butyl rubber may be used for lining large ponds, making it reasonable to consider the possibility of quite large slurry reservoirs which need only be emptied once or twice a year.

Considering a typical case of a 60-cow dairy herd, where all the slurry has to be handled, but every effort is made to avoid any dilution with washing water or rain, the amount produced in a winter of 150 days, based on 10 gallons (45 l) per head daily, is 90 000 gallons (410 m³). In this condition the slurry is just suitable for distribution by a good tanker. The minimum dilution practicable for pipe-line distribution is 1 : 1; but where plenty of water is available, at least 2 : 1 dilution is recommended, and 3 : 1 dilution is preferable. At this dilution the amount to be handled per season will be 360 000 gallons (1640 m³). The minimum size of storage tank for this level of production is about 18 000 gallons (82 m³). A tank of this size would have to be emptied monthly where the manure is kept undiluted, and weekly where there is 3 : 1 dilution. Such a tank made in concrete is likely to cost in the region of £800. A similar sum spent on a pit lined with butyl rubber should provide for a capacity of about 50 000 gallons (220 m³).

Costs of Slurry Disposal

As with so many farming jobs, the 'fixed' charges for the equipment usually account for a high proportion of slurry handling costs. Low cost therefore depends on either using low-priced equipment, or finding a way of spreading the cost of more expensive tackle. Low-priced equipment generally results in men spending an unnecessarily long time doing a job which is at best tedious, and at worst thoroughly objectionable; so there is everything to be said for finding a cheap way of using high-capacity equipment. On small farms this almost inevitably leads to shared use of mobile or partially mobile machinery, rather than fixed pipeline systems. Nevertheless, it should be noted that one of the cheapest methods is provided by a simple pipe-line system, powered by a tractor pump. Typical cost of the specialized equipment for such a system, with delivery pipe and two rainers, is about £1000. Pumping rate should be about 6000 gal (27 000 l) per hour. Such a pump is fully mobile and can easily serve several farms, thus reducing capital cost per farm. Provision of two rainers makes it easy to avoid over-manuring. As a general rule it is inadvisable to apply more than about 1 inch (2·5 cm) at a time, i.e. about 11 000 gallons (50 000 l) through a rainer which covers ½ acre (0·2 ha) at a setting; but in most conditions it is better to move when about ½ inch (1·25 cm) has been applied. The pipe-line system has the advantage of avoiding running a tractor and tanker over wet land; of needing less labour; and of permitting use of a convenient amount of water for washing down yards, etc.

Slurry can usually be handled effectively by a tanker which can not only spread at the rear, but can if necessary remain on a hard road and deliver the slurry at a distance through piping and a rainer. A 750-gallon (3400 l) vacuum-filled and pressure-emptied tanker, complete with a short length of pipe and rainer, costs about £1000. Similar tankers equipped with a positive (e.g. scroll-and-stator) pump can be used for effective agitation of manure in the storage tank and can operate against higher pressures, but equipment cost is higher (e.g. £1200 to £1500). The working rate of this size of spreader where direct spreading is possible and transport distance short can be about 2000 gal (9000 l) per hour.

There is an advantage in the shared use of such tankers, since their high transport capacity and fast filling and emptying can result in considerable time saving at a low fixed cost of equipment for each participant. By comparison, individual use of the cheapest type of tanker often results in higher cost and a spreading rate well below 1000 gal (4500 l) per hour.

For some farms where policy on liquid manure disposal is firmly established and conditions are suitable for spreading by rainers, it may be best to install a permanent system, with electrically operated mixing and pumping and underground mains to which the portable rainers can be easily attached. Such a system is likely to cost about £2000 for equipment, excluding the storage tank and installation. Typical application rate is about 3500 gal (16 000 l) per hour.

Where slurry can be effectively utilized it has a considerable manurial value which helps to justify expenditure on provision of facilities to ensure that it is applied when and where it will do most good. For example, the value of pig slurry per fattening animal per year is equal to about 20 Units of N, 3 Units of P and 8 Units of K. At Great House E.H.F. a total of 2 in (5 cm) applied at fortnightly intervals to grassland from October to March supplied about 1200 Units N, 200 Units P and 500 Units K per acre. This was very effective in grass production, though only as effective as about 25 per cent of these large quantities applied as mineral fertilizers. Much smaller amounts, e.g. 150 Units of Nitrogen applied in spring, are more effectively utilized, and can reach up to about 75 per cent efficiency compared with the elements in mineral fertilizers. Cow slurries are not so rich in nutrients but can also bring about substantial savings in fertilizer bills if applied in suitable quantity at the right time.

Fertilizer Distribution and Drilling

Many factors have to be taken into account in deciding on the most economic equipment and methods for fertilizer distribution. Many of these factors are inter-related, and should not be considered in isolation. Nevertheless, it is convenient, for the sake of simplicity, to consider in turn such questions as methods of broadcast fertilizer distribution, including handling problems; choice of drills; and comparison of broadcast fertilizer distribution with combine drilling, and fertilizer placement.

Before considering details, mention should be made of the advantages of a high rate of work. This is doubly important because distribution and drilling are often required at a time of peak demand for labour, and because timeliness in sowing crops often has marked effects on crop yields. The effect of sowing date on yield is well illustrated by the results obtained on five experimental husbandry farms on the best sowing date for winter wheat. The variety 'Capelle', which is reasonably adaptable to a wide range of sowing dates, was drilled over a number of seasons at set intervals from September to December. September sowing was usually too early, especially in seasons when the soil was warm. The drop in yield was up to 50 per cent. December was too late, sowing at Christmas usually resulting in about 20 per cent yield reduction. November was on average slightly too late, and the ideal was mid-October to the end of October. It is therefore highly desirable on a farm where winter wheat is grown to get the bulk of the work done between mid-October and mid-November. Similar considerations apply to many spring-sown crops, though here the penalty for late sowing may be more severe. For such reasons, working rate is often more important to the farmer than performance in man hours per acre; so though 'one-man' systems of distribution or drilling are usually most efficient, it is sometimes worth while to have a second man to help in filling the hopper, even though much of his time is wasted.

One other general objective which needs to be considered is evenness of fertilizer distribution. There can be little doubt that complete evenness is the ideal; but there is little precise information about the physical or economic effects of various degrees of lack of uniformity. Experience and experiments show that some kinds and degrees of lack of uniformity are tolerable, but it is certain that heavy over- or under-dressing in wide bands can lead to such troubles as laid corn crops, when the overall result can be a serious loss of potential yield.

Methods of Broadcast Fertilizer Distribution

There are two main types of broadcast fertilizer distributors, viz. those with a 'full-width' hopper and those which throw the fertilizer over a bout considerably wider than the machine itself. The latter type has been so much improved that such machines are now generally preferred, except on farms where the crops grown are mainly roots and vegetables. Reasons for the popularity of the spinning disc type include relative cheapness, simplicity of construction, ease of cleaning and a high rate of work when applying light top-dressings. Machines which distribute over a wide bout from a central hopper can easily be adapted to carry large loads, and so can be suited to adoption of bulk handling methods.

Fertilizer Handling

One of the most important factors governing working rate when distributing fertilizer is the efficiency of handling the fertilizer from the farmer's store into the distributor hopper. Often, more time is spent on this work than on the actual spreading operation. Moreover, many methods of handling the fertilizer in bags involve a great deal of heavy physical work. It is therefore worth while to study improved handling methods, both in bags and in bulk.

Factors leading to efficiency in handling bags of fertilizer naturally include avoidance of all unnecessary picking up and setting down. One of the simplest and best systems is to keep the fertilizer on a trailer which is moved along the headland as work proceeds, the distributor running to the trailer for re-filling. In very long fields, if a high application rate is required and the hopper is small, it may be desirable to pull the trailer across the centre of the field, thereby halving the effective bout length. This applies particularly to work such as potato planting, where a planter with fertilizer placement equipment is used.

Typical performance of a distributor with a hopper of 6 cwt (300 kg) capacity and a bout width of 20 ft (6 m) when spreading 3 cwt per acre (375 kg/ha) of granular fertilizer, is about 5–6 acres (2–2·5 ha) per hour, compared with a net working rate equivalent to about 12 acres (5 ha) per hour on continuous spreading at 5 m.p.h. (8 km/h).

One of the advantages of bulk handling is that it enables one man, working alone, to achieve an overall spreading rate appreciably faster than is normally possible when sacks are used. (See Table A.1(ii).) The most common method of storage is on a flat, dry floor, the heap being completely covered with polythene during the storage period. The spreader is usually filled direct by front loader or by a hydraulic shovel attached to the rear of the machine. A typical overall performance of a machine with a 30-cwt (1500 kg) hopper which runs back to the store a distance of up to 1 mile (1·6 km), when spreading 3 cwt per acre (375 kg/ha) of granular fertilizer, at a bout width of 20 ft (6 m), is about 9–10 acres per hour (3·5–4 ha/h); i.e. much faster than that of an otherwise comparable machine using sacks and a small hopper. (See Table A.1(ii).)

Liquid Versus Solid Fertilizers

Some fertilizers can be applied as solutions or suspensions, while nitrogen can also be applied as gaseous (anhydrous) ammonia, provided that suitable equipment is used to inject the liquefied gas into the soil at an adequate depth. Experiments have shown that there are few significant differences in crop yields between the same amount of nutrients applied as a solid or in solution. Choice of method can therefore be allowed to depend on the cost of getting the necessary nutrients applied. In some areas contractors are willing to supply and spread solid fertilizers for only a very little more than the cost of the same fertilizer delivered in bags to the farm by the agricultural merchant. This is made possible by such considerations as the saving in bags, and competition to secure the order for the fertilizer. In other areas, contractors who distribute fertilizer solutions quote low all-in costs for both supply and distribution. Fertilizer solutions are at present made up mainly by dissolving solid fertilizers in water; and so long as this situation prevails, it seems unlikely that they will be particularly cheap.

Anhydrous ammonia is basically a cheap form of nitrogen; and it can be cheaply transported over long distances once the costly high-pressure containers have been provided. Technical advantages include a relatively slow rate of nitrification and so of uptake by crop plants, and this can result in a beneficial lengthening of the period of response to the fertilizer. A disadvantage is the necessity to inject it into the soil to a minimum depth of about 4 in (10 cm), though the cultivation effect can often be beneficial and injection equipment can often be effectively combined with essential cultivation implements for arable crops. For many crops, including grassland, potatoes and root crops, one heavy dressing properly applied in spring can produce good responses lasting into late summer. Britain's moderate winter soil temperatures make it generally inadvisable to apply ammonia in autumn for a spring response, but the development of the use of additives which delay nitrification shows promise.

Complete NPK fertilizer systems can be based on anhydrous ammonia for heavy nitrogen dressings, supplemented by liquid spray applications for P and K.

There is little to choose between labour requirements for good liquid systems and the best systems of bulk handling for solid fertilizers. Solid fertilizer manufacturers tend to favour providing fertilizers in polythene bags, rather than loose in bulk; and various systems of handling the bags on pallets have been developed.

Combine Drilling or Broadcasting for Cereals

Many comparisons have been made between combine drilling at 7 in (18 cm) spacing and using a plain drill at various coulter spacings after broadcasting the fertilizer. Results depend mainly on the level of soil fertility, greater advantage being obtained from combine drilling where fertility is low. For example, in 20 experiments with barley in which 40 units of P_2O_5 per acre were applied, the mean advantage of combine drilling over broadcasting was 1·2 cwt of grain per acre (150 kg/ha) on soils which were low in phosphate; but there was no advantage from combine drilling the fertilizer on soils moderate or high in phosphate. In another series of experiments, combine drilling on average gave almost 1 cwt per acre (125 kg/ha) increased yield over broadcast fertilizer and 7-in (18 cm) drilling, but only ½ cwt per acre (63 kg/ha) increase over broadcast fertilizer and 5-in (13 cm) drilling. These results are fairly typical of several other recent trials. In general, it can be expected that combine drilling will give a worth-while return of about 1 cwt per acre (125 kg/ha) more where fertility is low; but on soils where narrow coulter spacing is practicable, 5-in (13 cm) spacing will give about ½ cwt per acre (63 kg/ha) more than 7-in (18 cm). Narrow coulter spacing tends to be more effective with barley than with wheat.

Average yields are, however, by no means the only consideration. There are considerable advantages in the 'once-over' technique of combine drilling on wet, heavy soils. Moreover, narrow coulter spacing is often quite impracticable on such soils, owing to the heavier draft, and difficulties with clods. Narrow coulter spacing and plain drilling therefore tend to be of most value on light soils in a high state of fertility. In these conditions, maximum advantage is obtained if a wide drill with a large-capacity hopper is used.

The effects of a large drill hopper are illustrated by the results of trials at an Experimental Husbandry Farm where a comparison of working rates was made between a typical combine drill and a plain corn drill equipped with a very large hopper. Rates of work in typical conditions are shown in Table 15. The working rates in Table 15 are higher than the comparable 'standard' figures in Table A.1(ii) owing to the favourable conditions in which the work was done. In this trial, man hours per acre with the large plain drill and separate distributor were 0·49 compared with 0·74 for the combine drilling.

N.I.A.E. tests on use of plain corn drills showed that on average overall rates of work obtained by 'Users' approximated to two-thirds of the spot rate for drills having small hoppers, but reached three quarters of the spot rate for drills with large hoppers. In the case of a corn drill with a 15 cwt (760 kg) hopper, it is practicable in some circumstances to fill the hopper at the farm and return for a re-fill after having done

Table 15. Rates of Drilling Barley and Compound Fertilizer

	1 Plate and flicker distributor	2 Plain drill, narrow coulter spacing	3 Combine drill 20 coulter 18 cm spacing	4 Combine drill as col. 3, used for seed only
Item			_Equipment_	
Number of men	2	1	2	1
Hopper capacity				
cwt	12	15	4 seed $5\frac{1}{2}$ fert	4
kg	660	760	204 ,, 280 ,,	204
Application rate				
cwt per acre	2	1·1	1·1 ,, 2 ,,	1·1
kg/ha	250	138	138 ,, 250 ,,	138
Working width				
ft	17·3	13	11·7	11·7
m	5·3	4	3·5	3·5
Average work rate				
acres/h	6·5	5·5	2·7	3·4
ha/h	2·6	2·2	1·1	1·4

some 2 hours work and completed about 12 acres (5 ha). This can result in a very high field efficiency.

In attempting to decide on the drill size required, account must be taken of the number of days when drilling is practicable. On the farm where the work referred to in Table 15 was carried out, records showed that on average there are 18 days between mid-March and the end of April when soil and weather conditions are suitable for seed-bed preparation and drilling. Allowing for time spent on cultivation at the beginning of each drilling period, it was estimated that the number of possible drilling days averages 11, with a range from 5 to 18. Fairly long days can be worked when necessary; so assuming 9 hours daily for 11 days there are about 100 hours available for spring drilling. On this basis, the potential acreages for single drills of the types used are about 250 acres (100 ha) for the spring period for the combine drill and about 620 acres (250 ha) for the plain drill. On smaller areas, there could be advantages for high-speed work in more timely completion of the job. Capital and operating costs of drills are usually of relatively small

importance compared with crop yields; but large drills cost in
the region of £1000; so there is advantage in arranging for
shared use of such machines where individual farm acreages
are inadequate to justify purchase. There are considerable
advantages in using special types of drills for particular crops,
rather than adjusting a multi-purpose drill to sow the full range
of crops grown.

An efficient root drill can economize in seed and result in
easier singling. An efficient grass-seed drill will assist in securing a rapid spread of sown grasses over the land, thus reducing
weed infestation and giving expensive seed mixtures a good
chance of producing a perfect sward. A good bean drill will put
beans down out of the way of the birds and will keep going in
really difficult conditions, where some types of drill would fail
to work.

A small drill will work at a rate of about $2\frac{1}{2}$ acres (1 ha) an
hour, so a single drill will, if necessary, do all the work on quite
large farms. The decision whether to buy special drills or not
therefore depends mainly on the cost of the drills and the
improvement in yields, etc., that their use will bring about. A
further consideration is the saving of time in adjusting coulters
and feed mechanisms.

Spaced Drilling for Root Crops

There are many advantages in using spacing drills for crops
such as sugar beet, swedes and turnips, and for a wide range of
vegetable crops. They include small savings in the cost of seed,
and often substantial savings in labour. A typical comparison
in costs for a crop such as swedes might be as follows:

	Cost	
	Per acre	Per hectare
Ordinary seed for non-specialized drill	2·5 lb, £1	2·8 kg, £2·50
Graded seed with single-seed drill	0·5 lb, £0·5	0·56 kg, £1·25
Hand thinning of continuous braird	£18	£45
Thinning after 1 in spacing	£12	£30
'Walk-through' after 6 in spacing	£3	£8

Purchase of a spacing drill can be economically justified on quite a small acreage. For example, a 4-row unit drill may cost about £360. The rate of depreciation on such drills is low. At the 12-year life estimate for little-used machines (Table A.2(i) Group 4), annual charges would be:

Depreciation		£30
Interest on capital	$\dfrac{8\% \text{ on } £360}{2}$	£14·40
Repairs etc. say		£4
		£48·40

Effective use on as little as 4 acres would repay such costs; but there are clearly advantages in arranging to use the drill at somewhere near its potential, so that it can be replaced by a more up-to-date machine within a reasonably short time.

Drilling Sugar Beet

The main trend is illustrated by progress with sugar beet. Substantial improvement demands use of pelleted seed, usually of the monogerm type, which results in about 95 per cent of single plants. The aim in good conditions is usually a 6 in (15 cm) spacing, and the main problem is providing seed-bed conditions which ensure a high germination. Where conditions are adverse, as they may be very early in the season, 3 in (7·5 cm) spacing may be chosen, and this usually means subsequent thinning with hand hoes, though electronically controlled thinners may be used successfully. Row spacing is usually about 18 in (45 cm) and the aim is about 30 000 plants per acre (75 000/ha).

Work at Brooms Barn experimental station has shown the importance of timeliness of drilling. The ideal sowing date, when soil conditions are suitable, is the middle to end of March: April drillings result in about 1 ton per acre ($2\frac{1}{2}$ ton/ha) lower yield with increasing loss towards the end of the month. The differences in returns for quite small differences in sowing date towards the end of the ideal time are substantial. Unfortunately, spaced drilling of root crops is not an operation that can normally be speeded up by driving faster, since the quality of work tends to fall appreciably at speeds above about

$2\frac{1}{2}$ m.p.h. (4 km/h). So where a small drill is inadequate to do the necessary work in about 10 working days, use of a wider drill is likely to be economic. Herbicide spraying is a normal procedure. Overall spraying of residual herbicides is common, but the more persistent residual herbicides tend to be expensive, and for these band spraying at the time of drilling is preferable.

Some herbicides need to be applied 2 to 3 days before drilling, and must be mixed with the soil within 30 minutes of application. Other pre-emergence herbicides can be mixed with the soil using specially designed units immediately ahead of the drill units.

Inter-row Hoeing

The problems of inter-row cultivation have changed with increasing use of herbicides for weed control. The days when overall hand hoeing could be practised have passed. Even where labour is available, use for such purposes is uneconomic and unnecessary. Use of herbicides, if well carried out, in conjunction with effective use of spacing drills and possibly mechanical thinners, can sometimes completely eliminate hand work.

The first hoeing of row crops is still often done by a two-man operated steerage hoe. One-man mid-mounted and forward-mounted hoes are available but have hitherto been little used. Their disadvantages may include difficulty in doing close work accurately; extra cost; and more time needed to fit the equipment to the tractor. Special-purpose toolbar frames can be more effective than general-purpose tractors for close hoeing; but on most farms it is hard to justify use of such specialized power units. In the past they have had insufficient power for most work other than row-crop drilling and hoeing. There is a tendency for one-man hoeing to be slower than steerage hoeing, e.g. 1 acre per hour overall compared with $1\frac{1}{2}$; but the extent of any such difference depends on ease of control of the hoe, and ability of the driver to see the work clearly and at the same time to be able to see the work ahead.

Potato Planting

Considerations involved in choice of method of potato planting include the need to secure efficient utilization of the fertilizer; planting at the best depth and spacing, with a suitable size of seed; ensuring that the tubers are planted in fine, unconsolidated soil; and ability to get the job done at high speed when the conditions are right. In addition, there may be overall management considerations, such as the need to employ casual labour at planting time in order to have it available at peak labour demand times later in the year.

Planting by machine is demonstrably cheaper than hand work by casual labour, which costs at least £5 per acre (£12·50/ha). Moreover, results of machine planting can be entirely satisfactory from a technical crop husbandry viewpoint. A disadvantage of machine work on a large farm is the limited output of a single planter compared with a large planting gang. A good gang, with well-organized transport of seed, can plant up to $\frac{3}{4}$ acre (0·3 ha) per day per gangworker when planting chitted seed, and a little more of un-chitted; but average hand planting performances are only about $\frac{1}{2}$ to two-thirds acre per day of un-chitted seed.

Planters may be divided into two main types, viz. hand-fed machines, in which each tuber has to be placed by hand on to a conveyor or directly into a chute; and automatic machines, in which the tubers are moved from a hopper by the conveyor; on one type, 'misses' by the primary conveyor are automatically corrected by a secondary feed mechanism; and the operator has only to refill the primary and secondary feed hoppers, and generally supervise the work of a 2-row, 3-row or 4-row machine. Working rates are given in Table A.1(iv). With a hand-fed machine having one operator per row, it is impossible at normal tuber spacings to achieve a forward speed of more than about $1\frac{1}{4}$ m.p.h. (2 km/h). In practice, speeds may be considerably lower. The automatic planter avoids this human limitation, and with suitable graded seed positive-feed types can work at a speed of about 3 m.p.h. (4·8 km/h). Machines with a less positive oscillating feed can work at double this speed. Thus, a 2-row automatic planter can work faster than a 4-row hand-fed machine. Unfortunately, automatic planters are not very suitable for use with ungraded seed which has been chitted in a glasshouse; but improvements in planters and in chitting in a controlled environment can result in a satisfactory performance where seed is well graded and has short, sturdy shoots. Though machine planting is cheaper than hand work and will inevitably displace it, the saving in cost is unimportant compared with factors such as crop yield. For example, the saving in cost of planting may be £2 per acre (£5/ha), whereas a difference in yield of only $\frac{1}{2}$ ton per acre can mean a difference in output and gross margin of nearer £10 per acre (£25/ha).

Table 16. Weight of Seed Potatoes needed for Various Spacings

For 28 in		Average weight of tubers				
(71 cm) rows	$1\frac{1}{2}$ oz	2 oz	$2\frac{1}{2}$ oz	42 g	57 g	70 g
Spacing in row			Weight of seed			
in cm		cwt per acre			kg per hectare	
12 30	16	21	26	2000	2640	3250
14 35	14	18	22	1750	2250	2660
16 40	12	16	20	1500	2000	2440
18 45	11	14	17	1380	1750	2120

Results to date of experiments at Terrington E.H.F. to determine the effect of uniformity of spacing on yield show that unevenness has little effect on yield up to a coefficient of variation of 40 per cent, but that there is an appreciable reduction in the yield of ware at a 60 per cent coefficient of variation. Good spacing by automatic planters is therefore a necessary feature.

Experiments on row spacings show that for maincrops there is much to be said for 36 in (90 cm). This can provide a good ridge of about 600 cm^2 (93 sq in) cross-sectional area from only a very shallow depth of tilth.

Cost of seed is an important item in the total cost of potato growing. It is necessary to calculate fairly accurately the amount needed, and then to ensure that the planter puts in the required quantity. The figures in Table 16 indicate quantities needed where the crop is grown in 28-inch (71 cm) rows and individual tubers average from $1\frac{1}{2}$ oz to $2\frac{1}{2}$ oz (42–70 g).

Potato Chitting

Sprouting potatoes before planting enables the farmer to ensure earliness of maturity without the risks involved in very early planting. It helps to ensure freedom from 'misses', and permits the control of early growth in such a way that the crop is very uniform. Careful control of the environment can be employed to produce the type of early growth required. For example, at one extreme it may be desired, in the case of an early crop, to produce one strong, dominant growth (called 'apical dominance') which results in a few potatoes of good size being produced very early in the growing season. At the other extreme, where a crop is being grown for seed, it may be desired to produce a large number of rather small potatoes, and this calls for a multiplicity of equally vigorous sprouts. The 'traditional' method of boxing the seed in chitting trays and stacking them in a glasshouse cannot provide sufficiently accurate control of temperature, and in any case requires a considerable amount of heat to guard against a severe frost. Moreover, it is impossible in practice to avoid considerable difference in light and temperature in different parts of the house, and this often leads to very undesirable differences in sprout growth.

Buildings such as brick barns can be fairly easily converted into artificially illuminated chitting houses, which with good management can result in uniform short green sprouts.

The lighting is produced by warm-white fluorescent tubes which are suspended from wires in such a way that the potatoes contained in stacks of ordinary chitting boxes receive more or less equal illumination. As with bulk storage of ware potatoes, almost any existing farm building can be made into a chitting shed, but it is an advantage if it has good brick walls and a good roof. It must also be possible to provide adequate ventilation when it is needed; but the chief alteration necessary is to insulate the walls if they are not of thick stone or brickwork, and to provide a good insulated covering either to the inside of the roof itself or to a ceiling erected between the store and the roof space.

A small amount of heating must be provided to protect against severe frost, and it is usual to install one or two 3 kW fan type space heaters – sufficient to give a temperature rise of about 20 °F (11 °C). These may be controlled by a rod type thermostat.

Standard wooden chitting trays 30 in × 18 in × 6 in (76 × 46 × 15 cm) are loaded at about 30 lb (14 kg) per tray and stacked in columns 3 ft (92 cm) thick with alleys between the columns. The lights, which can be obtained in lengths up to 10 ft (3 m), are suspended on wires so that they can be moved to a limited extent along the alleys; and the number of lights needed is one per 6–8 stacks of trays, i.e. 3–4 trays length on each side of it. Lighting is applied for 8–12 hours a day at off-peak periods, and the lights are moved along regularly so as to give equal illumination.

With main crops, natural dormancy usually lasts until about the end of December, and can be continued almost as long as desired by maintaining a temperature of 40–45 °F (5–7 °C). Thereafter, chitting can be started almost at will by raising the store temperature to 60–70 °F (16–21 °C). When once sprouting has begun, a temperature of 47–54 °F (8–12 °C) will produce a steady growth.

Different varieties require different environment regimes, so it is usually a mistake to attempt to house earlies and main-crops in the same store. During spring, when external temperature rises rapidly, it is often necessary to ventilate only during periods of lower temperature, in order to cool the store and so check excessive sprout growth. In some parts of the South West, where earlies are grown and winter minimum temperatures are fairly high, it may be impossible to secure adequate temperature control without using refrigeration equipment. A store in which temperature is to be fully controlled needs good insulation and effective air circulation; and such stores are inevitably considerably more expensive than the more common type. Where air is re-circulated – a necessary arrangement in a fully controlled store – it is necessary to keep a check on humidity as well as temperature, since too damp an atmosphere can cause excessive root growth.

A new glasshouse costs about £50–£60 per ton of potatoes, and a simple artificially lighted store costs about £25–£30 per ton if a building is available. There is no reason for changing from a good glasshouse to an artificially illuminated store, but if neither is available, artificial illumination is likely to be the better proposition, unless use of the glasshouse can be justified for purposes other than potato sprouting.

Annual cost with a converted building may be estimated approximately as follows:

	Per ton
	£
Depreciation on building and electrical equipment	3
Depreciation and repairs on trays and fittings	4
Running cost – heat, light and power	1
Total	£8

With a glasshouse, depreciation on the building will be higher. Running cost will depend on the heating method and season, but will be likely to average about £3 per ton. The cost of trays etc. will be similar. Total annual cost of using a glasshouse is therefore likely to be a little higher. In both cases there are additional labour costs, which are likely to amount to about £8 per ton. The return needed to balance the total extra cost of sprouting is therefore about £16–£20 per acre.

The return from sprouting is not easily measured. Apart from such factors as securing a good yield before blight occurs, there is the advantage of avoiding a late harvest. A week or two can make all the difference between success and failure with a potato harvester, and there can be no doubt that avoiding late-maturing crops is a substantial advantage.

Use of Transplanting Machines

Many crops, especially vegetables such as cabbages, broccoli and brussels sprouts, are grown in rows at fairly wide spacing within the row. Such crops can be fairly easily transplanted by a simple machine in which each row is planted by a single operator who places each plant either directly into the ground or into a simple device which transfers the plant to the furrow opened by the machine. Several observations both in Britain and in U.S.A. indicate that assuming the plants are carried on the machine in a handy position, and the machine is designed so that the planting operation itself is easy there is a maximum accurate planting rate of about 50 plants per worker per minute on straightforward work. This does not mean, however, that 3000 plants per hour will be set, since there are necessary stops for turning, and collecting a new supply of plants. In practice, one investigation gave an average planting rate of only about 28 plants per minute of overall working time.

However, if the best is to be obtained from a planting gang, it is necessary to aim at the rate of 50 plants per minute during actual operation, and this necessitates definite forward speeds which vary according to the spacing distance as shown in Table 17.

Table 17. Forward Speed to Plant 50 Plants per Minute at Various Spacings

Spacing Distance		Speed	
in	*cm*	*m.p.h.*	*km/h*
12	30	0·57	0·9
16	41	0·75	1·2
20	51	0·94	1·5
24	61	1·14	1·8
36	92	1·70	2·7

American work (on tobacco) indicates that there is nothing to be gained in accuracy by going slower than these speeds, but that any appreciable increase in speed can result in an unsatisfactory standard of work. For appreciably lower spacing distances than 12 in (30 cm) it is usually necessary to use a more complicated machine which enables 2 operators to feed the mechanism supplying each row.

An investigation in the East Midlands on autumn cauliflowers planted at 11 000 per acre (27 000/ha) with a 3-row machine operated by a gang of five showed that the job took 11 man hours per acre (27/ha) compared with 22 man hours per acre (54/ha) for hand planting by day work.

In this particular case the fifth man in the machine-planting gang handed plants to the three operators who actually planted them. This is not necessary if suitable trays are provided for holding the plants, so a reasonable objective for machine planting this crop is 9 man hours per acre (22/ha). The hand planting figure may also be lower where piece-work rates are paid. Typical performances by modern equipment are given in Table A.1(v).

In comparing the cost of hand and mechanical transplanting, it is necessary to consider machines with 2, 3 and 4 units. With a simple unit use of a 2-row machine operated by a gang of 3 is justified on 4 acres (1·6 ha) of autumn cauliflowers, and a 3-unit machine operated by 4 workers on 10 acres (4 ha). Moreover, the machine enables a small gang to get the job done more quickly. Machine planting is also often technically better than hand planting, owing to the mechanical firming of the soil on each side of the row.

Other ways of establishing crops which are usually transplanted include (1) drilling *in situ* and singling by hand hoe and (2) drilling to a stand with a single-seed drill. Both methods are used for all kinds of cabbages, broccoli and brussels sprouts.

Work at Luddington Experimental Horticulture Station has shown that spaced drilling can result in a considerable saving of labour compared with transplanting, especially for closely-spaced crops such as leeks. Direct drilling of autumn cauliflowers resulted in earlier maturity and higher yields.

Field Crop Spraying

Spraying for weed control became firmly established with the introduction of 'growth regulating' herbicides for spraying cereal crops. A very high proportion of corn crops is sprayed, mostly against easily-killed broad-leaved weeds by inexpensive chemicals such as MCPA or 2, 4–D, at a cost for chemical of about £1 per acre; some against more difficult weeds such as wild oats and black-grass, using expensive chemicals at a cost of chemical in the region of £5 per acre. The complexities of well-planned spray programmes are tending to increase. It is useless to attempt to kill weeds such as chickweed, polyganum, cleavers or mayweed with MCPA; and when difficult weed control problems arise there is much to be said for arranging for a specialist contractor to undertake the work. No attempt is therefore made here to go beyond a very brief review of economic aspects.

Before spraying corn crops became a general practice, it was reckoned that the average loss of yield in spring corn from weeds that are easily controllable was of the order of 10 per cent. Now that the land is generally much cleaner, the basic situation has changed, and there must inevitably come a time on land that is really well farmed when it becomes necessary to consider whether routine spraying every year shall continue. Factors in favour of continuing to spray include easier harvesting, and avoidance of rapid re-infestation from weed seeds. There is, unfortunately, no simple way of assessing the value of the benefits from spraying, but it is fairly easy to arrive at a rough estimate of the cost of the work.

While most weeds of cereals are now easily controlled by well-executed spray programmes some weeds, such as wild oats, are more difficult to keep in check especially where continuous cereal growing is practised. Work at Boxworth E.H.F. showed that a reasonable degree of control could be achieved by very late sowing of spring barley; but the resulting loss of yield averaged 0·4 ton/acre (1000 kg/ha), and control by herbicides, though costly, is more economic. Some chemicals, e.g. Avadex, have to be mixed with the soil before drilling; some have to be sprayed on when the wild oats are at a critical growth stage, usually the $1-2\frac{1}{2}$ leaf stage. Some newer herbicides can, however, be used at a much later stage of development of the weed, e.g. when its tillering has finished: and these are suitable for control in winter wheat.

Though the need to spray for weed control may decrease as land becomes cleaner, there are many other spraying operations which are likely to become more widely practised. Examples include spraying sugar beet with insecticide to prevent attacks of aphids, which not only do direct damage but also spread virus diseases – a job which may have to be done several times a year; spraying crops for similar purposes and also with a fungicide to protect against potato blight; and very varied spray programmes for crops such as peas and vegetables.

Choice of Spraying Machine

On some farms the choice of a sprayer is now a simple problem, for where weed control in corn crops is the only serious requirement there can be no doubt that the best investment is a cheap 'low-volume' machine costing about £100. On many farms it will be possible to write off such a machine in a single year and still show a profit on the spraying operation. Such a machine can only deal effectively with a limited range of materials, but it is usually possible to call in a contractor for those jobs for which the machine is unsuitable.

If the sprayer is needed for a wider range of jobs, including such operations as potato spraying, where a higher volume of material is needed to secure a good cover, a more expensive 'universal' type of sprayer is needed. Such machines may cost £200–£500 or more and should be capable of applying a range of sprays effectively at rates ranging from 5 to 100 gallons per acre (56–1120 l/ha).

With either the low-volume or universal type of sprayer there is a choice between models which are mounted on the tractor and those which are trailed. The mounted models need to be easily attached and removed, and if that is achieved this type has the advantage of compactness and suitability for use in row crops with narrow headlands, and also ease of transport. Some farmers need a machine which is suitable for either field or orchard spraying. This is unfortunately not as simple a matter as it was formerly, when high-volume hydraulic sprayers were considered satisfactory for all kinds of orchard work. Most fruitgrowers now wish to use low-volume 'air-flow' automatic sprayers, since they are labour-saving.

Contract Work

While recent developments have made the purchase and use of spraying machinery by farmers themselves much more attractive from both technical and economic standpoints, there still remain some tasks which can be most efficiently done by contractors. These include spraying which calls for the use of highly toxic compounds. Contractors may also be able to employ such expensive equipment as aircraft in circumstances where it is impossible or undesirable to get through the crop using ground equipment.

One of the many good reasons for employing a reliable contractor to do some types of spraying is the need for a thorough technical knowledge of all aspects of the work. Among the hazards which must be reckoned with are the possibility of damage to neighbouring crops as well as to that being sprayed, and also the risk of injury to employees who are not sufficiently careful in dealing with dangerous chemicals.

Cost of Spraying

A farmer can easily calculate his own costs at present price levels using the methods already explained. For example, if a low-volume sprayer will cost £100, and will be used to spray only 20 acres (8 ha) a year, annual depreciation (Table A.2 Group 5) will be

$$\frac{£100}{10} = £10,$$

and interest will be 8 per cent p.a. on £50 = £4. So these two items will represent

$$\frac{£14}{20} = £0 \cdot 70 \text{ per acre } (£1 \cdot 70/\text{ha}).$$

Maintenance and repairs should not be expensive, and operating cost (tractor and labour) need be little more than £0·40 per acre. In these circumstances, therefore, the farmer could operate his sprayer at a total cost (excluding materials) in the region of £1 per acre (£2·50/ha), of which depreciation represents half. With a more expensive high/low volume sprayer costing about £300, it will be necessary to cover about 50–60 acres (20–25 ha) a year if total operating costs are to be kept down to about £1 per acre (£2·50/ha), but there are circumstances where on smaller areas than this, use of such a sprayer can be cheaper or better than relying on a contractor. Operating costs fall with increasing annual use. As the use of spraying machines has developed, manufacturers have greatly improved the resistance of their machines to corrosion, and there is no reason why heavier depreciation rates than those indicated in Group 5 of Table A.2 should be necessary, provided that farmers take reasonable precautions in regard to such matters as washing out immediately after use, and handling only the range of chemicals for which the machine was designed.

Having calculated the cost of operating his own sprayer in this manner, farmers can compare their own costs with that of having a contractor in to do the job. Contract prices (excluding materials) are usually in the region of £1 per acre (£2·50/ha) for low volume machines, £1·60 (£4/ha) for medium volume and £2 per acre (£5/ha) for high volume work.

The advantage of being able to do the job when conditions are ideal cannot be estimated, but this is an important factor to be considered when deciding whether to buy a machine or to get the job done by a contractor. On the other hand, the warning given above concerning the dangers of insufficient technical knowledge or care should not be lightly dismissed.

Spraying by Aircraft

Aircraft have the advantage of being able to deal with crops such as brussels sprouts at a stage when some damage would be done by the passage of a tractor and sprayer through the field. Aircraft are also very suitable for spraying potatoes when crop and ground conditions are such that a tractor would do serious damage. The cost of work done by aircraft is inevitably high; and contractors who operate both aircraft and ground machines tend to choose the latter when time is not pressing and ground conditions are suitable. Special-purpose high-clearance tractors with wheel-shields have been developed for spraying tall row-crops.

The cost of spraying by aircraft naturally depends considerably on the area to be done and the suitability of the site for aircraft operation. Normal application rates are only 2–3 gal/acre (22–34 l/ha). Fixed-wing aircraft are generally cheaper to operate than helicopters, typical application costs being £1·50 per acre for fixed-wing machines and £1·75 for helicopters. Fixed-wing machines are more restricted in the type of work they can successfully carry out.

Weed Control in Peas

Developments in herbicides, especially the residual pre-emergence types, have quickly transformed the problem of weed control in peas and many other crops. Wild oats and many other difficult weeds can be effectively controlled. Peas were formerly commonly grown in fairly wide rows in order to facilitate some mechanical hoeing; but success with use of herbicides makes any form of hoeing unnecessary, and leads to drilling with narrow row-spacing. A typical cost for herbicide for spraying in peas is about £3 per acre (£7·5/ha), the spray being applied by either farmer or contractor. Spraying with pesticides is often necessary, and is often a contract operation. A typical contractor's charge, including material, is about £4 per acre (£10/ha).

Another pea-spraying operation which may be mentioned here is the use of a desiccant prior to combine harvesting of dried peas. This is particularly effective where crops are weedy: and a typical inclusive contractor's charge for it is £3–£4 per acre (£7·5–£10/ha).

Weed Control in Carrots

As with peas, there has been a fairly rapid change from hoeing to use of herbicides for weed control, especially in the case of carrots grown for processing or pre-packing where it is essential to regulate size of the roots by narrow row spacing. Though beds of narrow-spaced rows are ideal from the crop production aspect, mechanical harvesting of wide beds can be difficult, and this factor leads to patterns such as a band of three closely spaced rows with a wide spacing between these three rows and the next. Where such spacing is employed it often pays to do some tractor hoeing, and this in turn leads to band spraying to save cost on chemicals. A few contact herbicides are used, as well as a wide choice of both pre-emergence and post-emergence residual herbicides. One of these is paraffin (certain grades) which is used at about 60 gal/acre (670 l/ha) or less than half of this if band-sprayed. The degree of technical efficiency needed for successful band spraying must not, however, be overlooked. This is a job that normally has to be carefully done, 4–6 rows at a time, using an accurately steered outfit. With a wide tractor-mounted sprayer there is likely to be too much bouncing and swaying of the boom to give the required degree of accuracy, even if the drilling is straight and the joins are perfect. In any case, it is advisable to hoe while spraying, and this cannot usually be done over more than one width of the drill.

Spraying Against Potato Blight

Human consumption of potatoes in Britain averages almost 200 lb (90 kg) per head of population per year; and to supply this demand, almost 600 000 acres are grown. Crop yield fluctuates considerably, and an appreciably lower acreage would be adequate in most years. Yield tends to be higher in years of high summer rainfall; but wet weather in July and August may result in an infestation of potato blight. An early infestation reduces yield in spite of the favourable soil moisture conditions, while a later epidemic causes poor keeping qualities and difficulty in sorting for market. In the past, some farmers have followed a 'routine' spraying against blight, regardless of the weather.

Spraying is an expensive operation, and research has shown that in years when blight does not become epidemic early in the season, spraying can actually reduce yields owing to the mechanical damage done by tractor wheels, and the hardening of the leaves caused by some sprays. On the other hand, in areas such as the Fens in years when blight becomes epidemic at the time of corn harvest, effective spraying or dusting can secure a worth-while increase in weight and also a great improvement in keeping quality.

For most areas the best plan for a farmer to follow is to delay his first spraying until the Ministry of Agriculture gives a warning of a 'Beaumont period' or 'critical period', in which blight can spread rapidly. This first spray must be applied within a week of the warning, and repeated within about 2 weeks unless the weather has turned hot and dry in the meantime.

A disadvantage of relying on a contractor for spraying against blight is that in a season when blight is epidemic, operating conditions are always bad, and only the best contractors are able to keep to a schedule which will keep crops free from the disease. The object of the spraying is therefore not achieved, and blight may rapidly destroy the haulm. Recent technical developments make it easier for farmers to do their own blight spraying at a reasonable cost, and it seems likely that progress in this direction will continue.

In addition to such well-tried compounds as proprietary preparations of copper oxychloride and cuprous oxide, effective compounds of zinc and tin are available. Most may be applied at about 15–30 gallons per acre (170–340 l/ha) by medium-volume general-purpose sprayers.

The results of a series of trials extending over six consecutive years at Bridget's Experimental Husbandry Farm, based on the 'Weather' system already mentioned, showed that with the best sprays there was a substantial net return from spraying in four years out of six. Upwards of 2 tons per acre (5 ton/ha) extra yield of ware was obtained, with net financial returns of the order of £20–£40 per acre (£50–£100/ha). In two years out of six there was a small negative response, and cost of the 'insurance' was of the order of £10 per acre (£25/ha).

Irrigation

The capital and operating costs of providing irrigation, and the returns that can be expected in different conditions, vary widely. They are influenced not only by the type of water supply and the cropping, but also by the size of the scheme, the type of equipment chosen, the climate of the region, and the skill with which the equipment is used. It is easier to give average figures for costs than to predict returns, but the cost figures given below should only be used as a very approximate guide, since conditions and equipment differ considerably from farm to farm. Figures quoted make no allowance for extra costs that may be involved in obtaining water from a difficult supply, or in providing pumping installations with a number of automatic or semi-automatic controls. Nor is any allowance made for pumping to fields situated appreciably higher than the water supply. For simplicity, costs refer only to the type of equipment which employs lines of overlapping rotary sprinklers of small diameter.

For intensive vegetable growing, where all crops may need water at the same time, the cycle for one-inch applications of water should not exceed about 10 days, but many farmers growing sugar beet and potatoes may be satisfied if they can get round the irrigated area once a fortnight. The longer the cycle, the cheaper the capital cost per acre; but since the real value of irrigation equipment in Britain is experienced during droughts of only a few weeks duration, it is inadvisable for most farmers to reckon on a cycle of much longer than 10 days.

Capital Costs

It is convenient to consider capital costs in relation to the equipment needed to provide for a 1-acre (0·4 ha) setting. In addition to any costs incurred in making available a supply of water, it is usually necessary to provide a pump, portable distribution pipes, a few special joints and valves, and about 20 sprinklers. The greater the throw of the sprinklers, the fewer needed; but with a very few large-diameter sprinklers the size of droplets may be unsatisfactory.

For the cheapest portable installations, a tractor pump costs in the region of £400; 3-inch (7·5 cm) distribution pipes cost about £1 per yard (£1·1/m); and a sprinkler line complete with the necessary standpipes and sprinklers to cover 1 acre typically costs from about £350. So the cheapest possible installation capable of covering 1 acre (0·4 ha) costs from about £750 (e.g. Fig. 8 upper). This equipment is only capable of working adjacent to the water supply. In most circumstances additional equipment will be needed to reach land at a distance from the source of water. With additional piping and the fullest possible use of the equipment, it may be possible to make the equipment cover up to 4 acres a day, or 40 acres in a 10-day cycle. Generally, however, management difficulties will result in less than full use being made of the equipment.

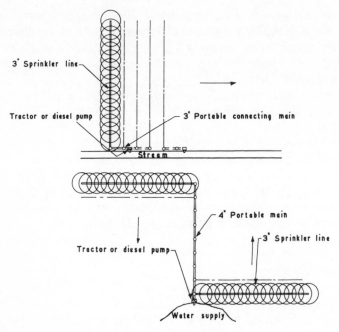

Fig. 8. Two simple layouts of irrigation sprinkler lines. (Mains, 102 mm; sprinkler lines, 76 mm.)

Surveys show that the average capital cost per acre setting, excluding source works, is in the region of £1000 per acre (£2500/ha); and on a basis of 40 acres (16 ha) covered per acre (0·4 ha) setting, this is a capital cost of £25 per acre (£63/ha) irrigated. As the practice of irrigation is extended, expenditure on source works of one kind or another tends to become necessary, and this may considerably increase total capital cost per acre irrigated.

In future, there will always be either a charge levied by river authorities or charges arising from the cost of providing storage, or a combination of the two. The cost of constructing reservoirs depends on the suitability of the site for adoption of cheap methods of construction. Where conditions are ideal for constructing a reservoir about 6 ft (2 m) deep simply by use of tractor-operated earth-moving equipment, the cost is likely to be in the region of £1000 per million gallons (4·5 million litres); and this quantity of water, in the absence of summer replenishment, is only sufficient to provide for about 44 acre inches (18 ha to a depth of 2·5 cm). Cost will be considerably higher where the reservoir has to be lined, or where there are other site difficulties. Thus, the total capital cost of a 1-acre setting can easily be £2000, £3000 or more; and the corresponding costs become £50, £75 or more per acre (£125, £187 or more per ha).

Operating Cost

The life of irrigation equipment and installations depends on the nature of the equipment and the conditions of use. For costing purposes some distinction must be made between permanent and portable equipment. Much of a permanent installation such as wells or reservoirs, electric motors and buried mains can safely be given an estimated life of 20 years, but the life of a pump and of a Diesel engine should not be expected to exceed 15 years, and may be less if the plant is fully used. If the pumping plant is used fully a more conservative figure of 10 years is suggested.

As regards portable equipment, aluminium alloy pipe is reckoned in U.S.A. to have a life of 15 years, but 10 years would be a safer figure. The life of sprinklers will clearly depend a great deal on use, but may be expected to be 8–10 years. If the cost of providing a water supply is excluded, the capital costs for a 1 acre (0·4 ha) setting may range from about £1000 for the cheapest all-portable equipment to about £3000 for a more typical scheme where about two-thirds of the total is spent on fixed equipment. If the equipment covers 40 acres (16 ha) annual fixed costs will range from about £4 to £8 per acre irrigated (£10 to £20 per ha). Surveys show that labour costs normally range from about 3 man hours per acre inch where a man is employed full-time on moving the laterals, to about half this where the work is a part-time job. A typical figure is 2 man hours per acre inch, costing about £1·60. Pumping costs vary considerably according to the type of installation. With an all-electric plant, electricity used may be expected to be about 40 units per acre inch, costing about £0·50. Where an engine or tractor is used, pumping costs are likely to be about £2 per acre inch. Thus, the total cost of irrigation depends to a very large extent on the fixed costs. Typical total figures are from £3 to £6 per acre inch on a fully used installation, but may often reach £5 to £8 per acre inch where annual use is as little as 2 acre inches.

Water provided by public undertakings typically costs from £0·25 per 1000 gallons. Though this may at first appear to be a high price to pay for irrigation water, it may in fact compare favourably with the costs arising from building a large reservoir.

Returns from Irrigation

The returns from irrigation in Britain naturally vary with the dryness of the season and with the use that the farmer makes of the water. The following figures indicate what has been achieved in suitable conditions, where irrigation is allied to good management.

Potatoes respond to irrigation with greater reliability than most other crops. Typical average yield increase is about 40 per cent, ranging from 2 tons per acre (5 ton/ha) for 2 in (5 cm) of water applied to earlies, to 3 tons per acre (7·5 ton/ha) for 3 in (7·5 cm) of water applied to maincrops. The value of this increased yield can be reckoned at about £50 per acre (£125/ha) for maincrops and often considerably more for earlies. Naturally, the average yield response is much greater on sandy soils in dry regions than it is in the traditional early potato growing areas of the West; but even in the West, irrigation may be worth while, since a small extra crop yield early in the season may be sold at very high prices. With some maincrops, such as King Edwards, it is usually inadvisable to irrigate before tuber swelling begins about mid-June, but earlier irrigation may benefit varieties which have less tendency to produce a large number of tubers, e.g. Majestic. In a wet year, very little response can be expected and applications may even be harmful. In a Nottinghamshire survey, average 'break-even' extra yields were 6 cwt per acre (750 kg/ha) of earlies and 8 cwt per acre (1000 kg/ha) of maincrops.

Sugar beet can be expected on average to give a positive

response to irrigation, but apart from exceptionally dry seasons, the return is much less than that for potatoes. A typical increase in yield is about 15 to 20 per cent, or 2 to $2\frac{1}{2}$ tons per acre (5000–6000 kg/ha), for the application of 2 in (5 cm) of water. Irrigation may be applied before drilling to a dry seed-bed, but it is usually unnecessary before the leaves meet in the rows, or after the end of August. In a Nottinghamshire survey, the average break-even increase in yield was 12 cwt per acre (1500 kg/ha). In very dry years on sandy soils, irrigation can make the difference between a complete failure and a full crop; but in a wet year there may be no response to irrigation.

Vining peas respond well to irrigation, with an average increase in yield of about 25 per cent or 8 cwt per acre (1000 kg/ha). In a dry season, an increased yield of 50 per cent may be obtained from a single irrigation of 1 in (2·5 cm) at flowering time.

Grassland responds best where adequate fertilizers are applied and the soil is never allowed to dry out. Where irrigation is based on a moisture deficit system, the deficit should not exceed $1–1\frac{1}{2}$ in (2·5–4 cm) on light land where the grass is closely grazed, or $1\frac{1}{2}$ to 2 in (4–5 cm) on heavy land.

Grass uses water at a rate of about 1 inch every 8 days in June and July, 1 inch every 10 days in May and August and 1 inch every 14 days in April and September. A typical seasonal increase in yield is 1 ton of dry matter per acre for about 5 in (12·5 cm) of water. The value of this depends on the use made of the herbage, but may range from £20 to £30 per acre (£50–£75/ha). Returns from milk production tend to be more certain than those from beef. Stabilization of yield, not only during the season but from year to year, is one of the advantages, since this permits full stocking and so promotes efficient grass utilization.

Irrigation for hay or silage production is more difficult to justify; but a small amount of irrigation, e.g. 2 in (5 cm), may result in an increased yield of about $\frac{1}{2}$ ton dry matter per acre (1·25 ton/ha).

As a general rule, irrigation equipment will be used for such purposes only if it cannot be used with advantage on crops of higher value.

Cereals. Irrigation of cereals is in the same category as

irrigation of grassland for conservation. If the equipment is free, a small amount of water in a dry season on light soils will produce a good return; but it is not normally worth while to purchase extra equipment specifically for this purpose.

Fruit and vegetable crops can give very profitable returns from irrigation. Here the returns are very variable; but as a general rule, market prices will be high when unirrigated crops suffer from lack of moisture. Typical responses where irrigation is used to meet an appreciable water deficit are:

| Crop | Quantity applied | | Returns | | |
	in	cm	cwt/acre	kg/ha	£ per acre
Bush fruit	4	10	16	2 000	100
Strawberries	2	5	12	1 500	70
Dessert apples	3	7·5	32	4 000	100
Cauliflower	2	5	60	7 500	150
Lettuce	2	5	80	10 000	160
Celery (self-blanching)	4	10	160	20 000	300

It should be emphasized that the figures given for fruit and vegetables are not averages, but are examples of the order of the returns that can be achieved by good management. The list of vegetable crops which can be profitably irrigated is a long one. In addition to the examples given, beans, carrots, beetroot and most kinds of leaf crops such as cabbages, spinach, etc. all respond to irrigation in a dry spell. These crops can justify irrigation even where water is expensive, and the scale or nature of operations makes capital cost per unit area high.

Trends in Irrigation Systems

Experience shows that low-cost irrigation systems can fail to give satisfaction because of the labour involved in moving portable pipes through wet crops. More expensive systems have therefore been developed, not merely to save labour, but also to eliminate the distasteful part of the job. Examples include:

Daily pipe-move system. More than one lateral is provided, and each has sprinklers sited at only every 2nd or 3rd position along it. The sprinklers may be moved first from point to point

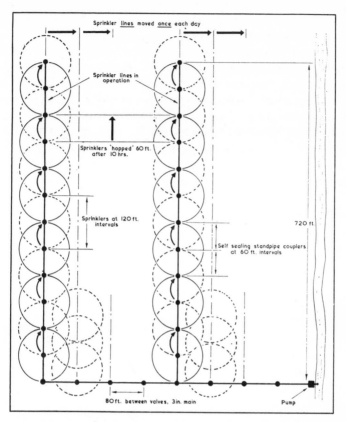

Fig. 9. Labour-saving method of sprinkler operation, using sprinklers with a long throw, in conjunction with a self-sealing bayonet-type standpipe coupler. With the layout shown, the 2-inch (5 cm) sprinkler lines are only moved once daily (20 hours application); but after 10 hours, each sprinkler is moved 60 ft (18 m) to the next standpipe coupler. With two lines of 720 ft (220 m) each, 2·6 acres (1·05 ha) are covered daily. The outfit can cover about 40 acres (16 ha) in a 15-day cycle.

(Wright-Rain)

along one lateral, and then over to the adjacent lateral (Fig. 9). The system allows a long interval between movement of laterals.

Seasonal 'Solid-set'. A complete system of small bore pipes is provided, and only the sprinklers have to be moved. Provision of a solid set makes it possible to provide the necessary sprinkler equipment to meet the requirements for protection of fruit crops from late frosts.

Mechanical-move systems. There are several systems of moving sprinklers over or through growing crops. In one system a light vehicle carrying sprinklers on a fixed boom is gradually driven by water power along the rows of crops such as potatoes. It automatically switches off the supply when the bout is completed, and then has to be moved to the next operating area. There are also large rotating-boom systems which are tractor-mounted, remain stationary during operation and can cover up to about 3 acres (1·2 ha) at a setting.

CHAPTER SEVEN

Harvesting and Conserving Green Fodder Crops

*Comparisons of Main Methods of Conservation. Haymaking.
Bale Handling. Silage Making. Silage Making Systems. Wilt-
ing for Silage. Making High-dry-matter Silage. Grass Drying.
Barn Hay Drying. Zero Grazing.*

Comparison of Main Methods of Conservation

Attempts have been made from time to time to compare, from
either technical or economic aspects, the main methods of
conserving green crops, viz. outdoor haymaking, silage making, grass drying, and sometimes barn hay drying. This is not a
particularly profitable exercise, since three of the four
processes are essentially different, in that the crops are normally cut at different stages of growth, and so inevitably
produce fodders of differing chemical composition and feeding
value, even assuming that the processes themselves are effectively applied. So far as an individual farm is concerned, the
aims in fodder conservation may be very varied. At one end of
the scale, the object may be to produce a bulky fodder (hay or
silage) as cheaply as possible, while at the other extreme it may
pay to spend a good deal on producing a feed that is rich in
protein and starch equivalent, and contains a minimum of fibre.
The latter, if preserved as high-quality dried grass or dried
lucerne in the form of meal or cubes may be worth above £60 a
ton, whereas the older material, even if very nicely made into
hay, will probably be worth less than half that price.

Production of hay, at about 7 million tons annually in
England and Wales, accounts for about 70 per cent of all

conserved forage. Approximately 10 million tons of silage is made: but because of the lower average dry matter percentage this represents only some 30 per cent of total conservation. Dried grass production is usually less than 1 per cent of the total quantity.

The considerable developments in techniques of silage making have been accompanied by important advances in haymaking machinery, with the result that hay remains by far the most widely practised of the conservation methods. This is partly related to the fact, referred to in Chapter 10, that a high proportion of the country's dairy cows are still housed in cowsheds, in spite of the well-known advantages of loose housing. Work at Bridget's Experimental Husbandry Farm (Hampshire) is typical of comparisons between hay and wilted silage. The hay was cut by 5 ft (1·5 m) mower and crimper, usually followed by a second crimping; a 2-row tedder as required; a finger-wheel machine for windrowing; pick-up baler; a tractor-mounted swinging-arm bale loader, with high-sided tumble-loaded trailers; and an elevator-conveyor to deliver to the barn, where drying was by a diesel engine-driven mobile fan. The silage was cut by 5 ft (1·5 m) mower; wilted up to 24 hours to produce a dry matter of 25–35 per cent; picked up by double-chop forage harvester; carted in tipping trailers; ensiled by buckrake in a barn; and covered by a 200-gauge polythene sheet, well tucked in. Material was cut from the same field at the same time (usually in May) for barn-dried hay and silage. Hay cost about 30 per cent more to produce and was also more demanding in labour and management. However, total losses in making the barn-dried hay were lower (22 per cent compared with 32 per cent for silage) and total cost per pound of starch equivalent fed was approximately the same for both feeds. Similar trials in an area of lower rainfall at Drayton Experimental Husbandry Farm (Warwickshire) resulted in inclusive costs of growing, harvesting and storage per ton of dry matter for barn-dried hay about 40 per cent higher than those for wilted silage; but the barn-dried hay produced 23 per cent more edible dry matter per unit area. (The comparable figure for Bridget's E.H.F. was 17 per cent higher yield from hay.) In trials in a high rainfall area at Great House E.H.F. (Lancashire) where a batch type hay drier was used, barn-dried

Table 18. Average Losses in Making Barn-dried Hay and Wilted Silage at Great House (Lancs.) E.H.F.

	Barn-dried Hay % loss	Wilted Silage % loss
Loss of dry matter in field	15·1	7·6
Loss of dry matter on drier	5·1	—
Loss of dry matter in store	3·3	20·9
Total loss of dry matter	23·5	28·5
Average digestibility %	68·2	68·5

hay produced only 5–10 per cent more edible dry matter, and distribution of losses was as shown in Table 18. The smaller advantage for barn-dried hay was due to 15 per cent field losses, and indicates the greater difficulty of drying down to a moisture content suitable for barn drying.

The variations between the averages of these individual farms, and even greater variations due to season, emphasize the need to be cautious in studying the results of any comparisons made on a single field or farm in one season.

The value of any feed in practice naturally depends on the use that the animals can make of it; and this can be an important consideration in deciding on a conservation system. For example, cattle which are offered barn-dried hay tend to eat more food than they would if fed even good wilted silage. High feed intake is usually an advantage for beef production, but not necessarily so for milk production, since the main effect may be that the cows weigh more at the end of the winter. This does not mean that barn-dried hay is unsuitable for milk production, but it cannot be economically fed *ad lib.* as ordinary silage can. This can be an important consideration on large farms with loose housing, but is no problem on small farms with cow sheds, where the hay is hand fed in any case. Thus, barn-dried hay tends to be more attractive where cow-sheds are used, than for large-scale dairying based on loose housing.

Haymaking

The essence of successful haymaking is to speed up the process in good weather and get the hay cured and collected within the shortest possible time after cutting. Shortening the period of exposure to the weather usually results in appreciable improvement in quality of the hay, and any such improvement will pay for a good deal of extra time and trouble in the making, and also for a considerable capital investment in equipment. At this job, 'man hours per ton', though important as the measure of what a limited labour force can do, is sometimes a less useful yardstick than 'tons per hour': efficient mechanization requires both high speed and a great output from a limited labour force. Cost per ton is usually low and relatively unimportant compared with the potential value of the product. At the one extreme, very bad hay may be quite worthless apart from some slight value as a source of manure: at the other end of the scale, really good hay can be equivalent in value to the lower grades of dried grass and may be worth up to £50 a ton. The cost of making and collecting need seldom exceed £20 a ton, even if a great deal of machinery expense and labour are involved.

Choice of Haymaking Equipment
Now that the value of a fairly severe swath treatment at or soon after mowing is generally accepted, the question whether a finger-bar mower is the best machine for cutting can hardly be avoided. Flail mowers have many advantages, including ability to keep on working in difficult conditions, where finger-bar mowers repeatedly block. Flail mowers also carry out a fairly effective swath treatment as they cut, and so avoid the need to buy a machine such as a crimper or roller crusher. The cost of a flail mower may be approximately equal to the combined cost of a mounted finger-bar mower and a simple roller crusher, but should be lower if produced in comparable numbers. Rate of work with the flail mower is similar to that of carrying out the combined cutting and swath treatment operation in easy conditions, but is likely to be higher in crops where the work is difficult. It is reasonable to expect a 5 ft (1·5 m) cut flail mower to do 2 acres (0·8 ha) per hour (see Table A.1(iii)); and in

reasonable conditions there is little to go wrong. Cutting thick-bottomed grassy crops by flail inevitably requires a considerable amount of power. A 5 ft (1·5 m) flail mower needs a tractor of about 50 h.p. in such conditions. A 5 ft (1·5 m) swath has several advantages; it suits most existing haymaking equipment; and wider swaths tend to be slow in drying. There is, however, an advantage in a 6 ft (1·8 m) mower for silage making in crops of up to about 10 tons per acre (25 ton/ha) as cut.

Rotary mowers of multiple vertical spindle type are similar to flail mowers as regards ability to keep going in difficult conditions. Power requirement is fairly high. The initial swath treatment is not severe, and the crop is left in good condition for repeated use of a tedder.

The severity of treatment by crimpers and crushers depends not only on the machine chosen, but also on the type of crop, thickness of the swath, and adjustment of pressure on the rollers. Generally, a roller crusher is less likely than a crimper to cause severe leaf loss in leguminous crops; but a crimper can be more effective in ensuring an adequate first treatment in thick, grassy crops. Avoidance of blockages, and ease of clearing them are important features. Machines with rollers of reasonably large diameter appear less prone to blockage than those with very small rollers.

With tedders, a high rate of work is essential. In fine weather, on farms where reliance is mainly on tedding, it may be desirable to use the machine up to 3–4 times daily to secure fast drying.

A very high proportion of the hay made in Britain is dealt with by pick-up baler, and there is no present indication that any other system will supersede it in the near future. Hay wafering accompanied by barn drying could completely change the situation at some time in the future; but though wafering has now been in commercial use for several years in some dry regions of U.S.A., notably California, attempts to develop machines suitable for humid regions have not yet had much success.

Choice of pick-up baler on many farms depends on use of the machine for collecting straw after the combine, as well as for haymaking. The most common type of automatic ram type baler makes bales with a section 18 in × 14 in (45 cm × 35 cm) or 18 in × 16 in (45 cm × 40 cm) and of a length that may be variable from about 24 in to 42 in (60 cm to 105 cm). The bales are 'sliced' on one side as the hay enters the chamber, and are usually tied by two bands of twine running lengthwise. Automatic wire-tying machines are also available but most farmers prefer to use twine. The cheapest p.t.o.-driven machine costs in the region of £900, so purchase cannot be lightly decided. The advantage of the ram type is that the rectangular compact bales are easy to handle and to stack. Some farmers prefer to use an independent engine on the baler, but on most farms this is an entirely unnecessary luxury now that tractors are so frequently powered by engines of at least 45 h.p., have a wide range of forward speeds and an independent p.t.o. There are, of course, conditions on hilly land and with light tractors where engine drive is advantageous. Contract charges are usually about 5p per bale including twine, but farm costs where the machine is used a good deal are appreciably lower.

Work rates of balers vary widely with working conditions. The following figures indicate potential and typical work rates of medium-size balers, which normally have a bale chamber 14 in × 18 in (35 cm × 45 cm) in section, and make bales up to about 42 in (1·07 m) long.

	In hay	In straw
Potential work rate		
tons per hour	15	10
bales per hour	675	550
Typical overall work rate		
tons per hour	7	6
bales per hour	315	330

Larger balers of the common ram type are available. These tend to have a bale chamber of larger cross section, and are suited to farms where hay and straw are sold off the farm. Output of these larger machines compared with the medium size is often disappointing. This is partly because it is usually impracticable to form very dense windrows, and there is a limit to the forward speed at which the machine can be safely and effectively worked.

Low-density press-type balers are usually cheaper than the ram type, and are in some respect suited to haymaking on small farms. They are widely used in some Continental countries, but most British farmers prefer the neat and denser bales produced by ram type machines, especially where straw has also to be handled.

Roll-type balers had the merit of producing bales which were very resistant to rain if left in the field; but handling difficulties prevented wide adoption. However, half-ton roll bales have now been introduced from N. America.

Handling of hay and straw in big bales of approximately $\frac{1}{2}$-ton size offers the possibility of very fast baling and carting by a small gang or by only one man. This can lead to avoidance of losses due to exposure of hay crops to rain. However, economic exploitation of the potential benefits is likely to call for effective barn-drying or tunnel-drying systems, or use of chemical preservatives, except in very good weather. Good systems will also be needed for handling big bales to points of use, and particularly for feeding.

Cost of Operating a Pick-up Baler

The cost of operating a pick-up baler may be estimated, using data from Tables A.1, A.2 and A.3. In Table 19 it is assumed that the machine is a p.t.o. model costing £1000, working rate is $2\frac{1}{2}$ acres (1 ha) per hour, and that one man only is employed on the baling operation.

Bale Handling

Sound methods of handling bales from the field to store are no less important than the baling operation itself. A decision has first to be made as to whether the chosen system shall be designed to permit some further field drying. Some farmers wish to leave the bales, especially straw bales, in the field for a few days before carting when weather permits. This is a risky procedure, but if further field drying is considered a desirable

Table 19. Cost of Operating Pick-up Baler
Price £1000. Rate of Work 2½ Acres (1 ha) per
Hour

	Heavy use	Normal use	Light use
Annual use, acres (ha)	250 (100)	125 (50)	50 (20)
Annual use, hours	100	50	20
Assumed life, years	5	7	10
Annual fixed cost £			
depreciation	200	143	100
interest	40	40	40
Total annual fixed cost £	240	183	140
Annual running cost £			
repairs	50	30	12
tractor	60	30	12
labour	80	40	16
Total annual cost £	430	283	180
Cost per acre excluding twine £	1·72	2·26	3·6
Cost per hectare excluding twine £	4·30	5·65	9·0

Cost of twine approx. 50p per ton.

option, this will influence choice towards a system in which the bales are grouped into regular heaps of 8 or more before loading. If a single-bale loading system is preferred, this will normally make it necessary to keep well up with the baler, so that few bales are left out if rain comes.

Single-bale handling systems include bale throwers, which necessitate drawing a trailer behind the baler. If the baler is to be kept busy, a gang of at least two tractors and three men is necessary. With such a team the two large trailers must be tippers. One man will be fully employed in ferrying loads to the store, and the third will be kept very busy loading bales on to an elevator-conveyor, which distributes them in 'tumble-stacked' form in the barn. A hold-up with any of the three operations brings the baler to a standstill, and it is not easy to maintain a high output. More usually there will be a gang of two or three stacking bales into store, and no less likelihood that the baler will be held up. At a long transport distance it is essential to use three trailers. Most bale thrower systems require the use of short bales, not over 24 in (0·6 m) long.

Single-bale pick-up elevators which necessitate a man loading the moving trailer in order to build a reasonable load are unattractive. This involves hard work, and also some danger. Systems likely to succeed are those which do not require a man to work on the trailer.

When tipping trailers are used without a man loading, short bales should be made, and the load tipped at the foot of an elevator-conveyor. This system suits tumble-stacking in the store, and enables a gang of only two, or one man with part-time help at the store, to handle the bales at a good rate. Tumble stacking in a barn fitted with suitable bale-retaining sides has been found to provide a storage capacity of about 90 per cent of that obtained by hand stacking. It is occasionally necessary to go up and push a few bales towards the corners, and to adjust the conveyor discharge point. Feeding bales from a tumble-stacked store does not present any serious difficulties.

Handling Bales in Groups

Most handling systems involve assembling the bales into regular groups. The oldest way of doing this, by means of a manned sledge, has many disadvantages, and was rapidly superseded. Using a manned sledge not only reduces working rate of the baler. The man who is employed in stacking the bales can work considerably faster when building heaps from a windrow of bales dropped on the ground. There are therefore good reasons for the popularity of unmanned bale windrowers. In the long run, however, it seems inevitable that simple wind-rowers and hand grouping will be superseded by various types of automatic collectors. Simple 'automatic' sledges produce rectangular groups of 8, suitable for picking up and loading on to a trailer by means of a squeeze type front loader. There are also more complex types of trailer collectors which arrange the bales in flat groups ready for picking up by a specialized fork-type loader.

When loading trailers in the field with groups of bales, there is much to be said for avoiding manual load-building on the trailer. The most suitable trailer for such work is likely to be one with a large bed and fairly high bale-retaining 'ladders' at front, rear and one side. A false floor which tilts the heaps of bales slightly towards the fixed side is useful. The open side can finally be roped after loading, if necessary.

One type of bale-grouping machine is towed behind the baler, and groups 20 bales into a 5 × 4 package. Each group is automatically tied by two strands of heavy-duty polypropylene twine. The group is then released and subsequent handling and stacking is usually by means of a grip-type loader. An advantage claimed for the system is that retaining bales of normal size facilitates feeding or other final use.

For short transport distances where travel along public roads can be avoided, front-and-rear tractor-mounted carriers can provide an efficient handling system. A convenient load size is usually 16 or 20 bales at both front and rear. Many types of carriers are available, the simplest being a short-tined buckrake with bale-retaining back and sides, and the most effective being more sophisticated devices with retention of the bales by means of hydraulic squeeze action.

Bale Handling: Working Rates

Bale handling may be carried out by many combinations of machines and methods; and as might be expected, there is in practice a wide range in performance achieved by similar methods. The figures here discussed are based on published work study investigations by the National Institute of Agricultural Engineering and survey-type studies of performances on commercial farms carried out by the National Agricultural Advisory Service. Typical bale weights are 56 lb (25 kg) in hay and 35 lb (16 kg) in dry straw. Where a single 'compromise' figure is used, as in preparing Table 20, 45 lb (20 kg) per bale (50 bales per ton) is assumed. Where manual work is involved, the rate of work in bales per hour is likely to be lower in hay than in straw. The work-study figures show attainable targets; but the extent to which performance is likely to fall short of these targets in practice is indicated by the survey figures. For the whole job of baling and getting bales from field to store, a reasonable target is 1 ton per man hour for systems included in Table 20. Big-bale systems are potentially much faster.

As with all farm transport jobs, efficiency in moving bales from field to farm depends on load size, distance to be

Table 20. Outputs of Bale Handling Equipment

Type of Equipment or Job	Minimum No. in Gang	Working Rates bales/hour		Survey Man Min. per Ton
		Work Study	Survey	
A. Baling Only				
1. Random dropping of bales	1	—	370	8
B. Baling and Arranging				
2. Manned sledge	2	212–357	204	32
3. Unmanned collector. Vertical 8 heaps made by hand from windrow	1 + 1	480	342	19*
4. 'Automatic' sledge (8 bales)	1	—	312	10*
5. Bale accumulator (flat 8)	1	—	300	10*
C. Baling and Loading				
6. Bale thrower	1	203 †	250	18
D. Loading				
7. Single-bale elevator	2	298	250	24
8. Single-bale swinging arm loader (tumble loading)	1	420	190	16
9. Squeeze loader (heaps from jobs 3 and 4)	1	340–544 ‡	190	16
10. Squeeze loader or buckrake (heaps from jobs 2 and 3)	2	—	250	24
11. Impaler loader. Heaps from job 5	1	456–535 ‡	250	12
12. Specialized tractor-mounted bale carriers. Manoeuvring and picking up only	1	—	500	6
E. Unloading and Stacking				
13. Elevator from trailer	1 + 2	518	375	25
14. Elevator from random load tipped on ground	1 + 2	408	—	—
15. By elevator-conveyor. Tumble stacked	1	—	200	15
16. Stack in barn. Grip type loader from trailer	1	—	250	12
17. Stack in barn. Impaler loader from trailer	1	392	300	10
F. Pick-up, Transport and Stack				
18. Using front loader and carrier to pick up, transport 440 yd (400 m) and stack in barn	1	—	200	15

* Includes baling for comparison with job 2.   † 29 per cent of time non-productive.
‡ Trailer loaded without moving.

travelled and the speed of travel. Investigations show that where transport takes place over a long distance, average speed tends to be higher than where journeys are short. This is doubtless due to the fact that longer journeys usually include a higher proportion of time spent on good roads. Table 21 shows how transport time on bale handling varied with transport distance in a national survey.

Table 21. Relationship Between Transport Distance and Transport Time for Carting Bales

Transport Distance One Way		Transport Time for Double Journey
Miles	km	Min
$\frac{1}{8}$	0·2	3
$\frac{1}{4}$	0·4	6
$\frac{1}{2}$	0·8	10
$\frac{3}{4}$	1·2	12
1	1·6	15
$1\frac{1}{2}$	2·4	20
2	3·2	25

The cost of bale handling equipment is modest compared with the benefits from its use, except in the case of the most complex and specialized equipment. In some of the principal haymaking areas of U.S.A., much more expensive mechanical collectors and stackers are sometimes used. These employ automatic hydraulic devices to build a load on a trailer or self-propelled vehicle; and the complete load is then deposited by up-ending the trailer bed. These machines are designed for handling very dense bales, and even with these, the stacks are often very unstable. There is little likelihood that such machines could be economically used in Britain.

Handling bales from store to the stock is briefly discussed in Chapter 10.

Silage Making

The main attractions of silage making to farmers who practise it regularly on a large scale are (a) there is almost a certainty of being able to complete the job more or less on schedule in any reasonable weather conditions, and (*b*) all of the tasks can be effectively mechanized. Progress in mechanization is shown by the averages (Table 22) from national survey samples.

Table 22. Changes in Labour Used to Make Silage

Year	Silage Making Systems	Time cost man min per ton
1948	Mainly green-crop loader and hand forking	149
1958	Mainly forage harvester, with tipping trailer and buckrake	36
1965	Mower, full-chop forage harvester, forage blower, making wilted silage	26

With a modern flail forage harvester, a typical potential performance on straightforward direct-cut work in a heavy crop is a loading rate of about 10 tons per hour, absorbing about 20 h.p. at the p.t.o. in dry conditions. In worse conditions, specific output (tons per hour per h.p.) may fall to 0·3. In ideal crop conditions it can rise to 0·8, but such a performance is usually associated with very little chopping or laceration, and so is not usually aimed at, since a good type of fermentation cannot be ensured without a reasonably short chop. A good specific output calls for a sufficiently high forward speed and avoidance of excessive rotor speed.

Towing a full trailer up a 1 in 10 slope at 3 m.p.h. (5 km/h) will typically add about 7 h.p. to the power requirement; and in practice, a 45 h.p. tractor cannot be expected to exceed about 10 tons/hour loading rate.

Silage-making Systems

The organization of silage making depends on basic objectives, as well as on labour and machinery to be used. Making silage by a direct-cut system with a flail forage harvester is a relatively simple task. Many farmers are satisfied with this basic method, and are concerned primarily with finding ways of carrying out the work more and more effectively. Others, however, seek to improve the product by making a better-fermented silage of higher dry-matter content; and some wish to weigh up the pros and cons of aiming at high-dry matter silage stored in a tower and mechanically fed. These problems are inter-related; but the whole subject is so broad that it is considered more useful to discuss some of the mechanization management factors separately.

One general point cannot be too strongly emphasized. A high rate of work is essential to success; firstly, because in terms of length of silage-making season, crops must be ensiled at near the time when they are at their best, and not after they have become excessively fibrous and indigestible; and secondly, in the shorter term, the work must be done when the weather is reasonably fine, since good silage cannot be made from over-wet crops.

Choice of system depends primarily on the number of men available and the rate of work required. As a general rule, there is nothing to be said in favour of large gangs, since these tend to be inefficient in terms of output per man, and two small gangs working independently will usually achieve a higher output per day. The emphasis must therefore be on gangs of no more than four men.

Direct-cut One-man System

One man can make silage effectively using either an in-line forage harvester which remains connected to the tractor and is taken to the silo with the loads; or a side-mounted harvester which is easily disconnected and left in the field when the full trailer is taken to the silo. The silo should be sited so that several loads can be tipped near it. Ensiling is then done after a spell of cutting and transport. The working rate that can be expected from a one-man system at a short transport distance is about 0·7 acre (0·3 ha) or 5 tons per hour on loading and transport only. This is reduced to about 0·4 acre (0·16 ha) or 3 tons per hour when allowance is made for the time spent in buckraking and changing jobs. It is therefore possible to ensile about 24 tons in an 8-hour day.

Another one-man system involves the use of a moving-bed self-emptying trailer and a forage blower at the silo. The trailer is filled by a side-mounted harvester which is left in the field. The trailer unloading mechanism is operated by the p.t.o., and the material delivered by a cross conveyor at the rear is delivered on to the silo by a blower, which is also p.t.o. driven. This equipment is more complex and expensive than a tipping trailer and buckrake; but it is capable of ensiling the crop safely and in good condition in tall or inaccessible silos, where the buckrake method cannot be used. Working rate is similar to that given by the tipping trailer/buckrake ensiling system.

Two-man Direct-cut Systems
The simplest type of two-man system operates like the one-man tipping trailer/buckrake systems already described, but allows for an extra man and tractor to do the buckraking as the loads are brought in. In practice, buckraking can often with advantage be a part-time job. (In good conditions a man with a buckrake can ensile over 10 tons per hour; but a more usual rate is 5–8 tons per hour.) Thus, for example, it may be practicable for the forage harvester to work a long day, while the buckraking tractor works between milkings and at the end of the day. On this basis it should be possible to ensile 35–40 tons per day.

Attempts to keep both men fully occupied by arranging for the buckrake operator to help with transport are not usually worth while. A second trailer and third tractor are required, and hitching and unhitching may be troublesome. For high-speed work, a system using three men and two forage harvesters may be better.

Three-man Direct-cut Systems
Three-man teams are common, partly because it is relatively easy to arrange a system in which one man with tractor, forage harvester and trailer does the cutting and loading; a second, with tractor and trailer does the transport; and a third, with tractor and buckrake does the ensiling. A typical overall loading rate, allowing for changing of trailers, will be about 6 tons or 3 loads per hour. Subtracting 5 minutes per load for hitching and tipping, there are 15 minutes for transport; and at an average transport speed of 8 m.p.h. (13 km/h) this allows for travel in both directions at a distance of 1 mile (1·6 km). At shorter transport distances the transport driver may have time to spare, but this will depend on the actual loading rate. In good

working conditions the buckrake operator should have time to do the job safely and well. Output from such a team should be in the region of 50 tons per day.

Where the silo is easy to fill and is located very near to the crop, e.g. in the same field, it will be better to use two forage harvesters and have both run direct to the silo, without disconnecting either trailer or forage harvester. The two will keep the buckrake operator busy, and the team of three should do in the region of 10 tons per hour, compared with 6 tons per hour when only one harvester is used. A simple forage harvester is not an expensive machine, and the higher rate of work can easily justify the extra capital investment on a farm where silage quality is important. With this arrangement, ensiling rate can reach 70–80 tons per day; and because of buckraking problems it may be best in practice to work on two silos.

Four-man Direct-cut Systems
From what has been said already, it should be clear that the only valid reasons for four-man direct-cut systems are: a very heavy crop and high loading rate, coupled with a long transport distance; or the need to provide additional help for ensiling where two forage harvesters are used. As a general rule, it is a mistake to assume that there ought to be a fourth man using a hand fork at the silo. Silage quality is determined mainly by the quality of material put in, the processing of this material in the forage harvester, the rate of ensiling, and the effectiveness of sealing, rather than by any teasing out of material that can be done by a man using a hand fork. Extra care and skill in operating the buckrake are much more important; and the only valid reason for the additional man is any help that he can give to ensure the buckrake operator's safety when the work has become dangerous.

Wilting for Silage
One of the main basic problems in silage making is to decide what type of silage to aim at. The product can only be as good as the material ensiled; and for high quality, cutting of a young crop is essential. But young crops always contain only a low proportion of dry matter, and are difficult to make into good silage if ensiled direct. The object of wilting is to remove some

of this moisture in the swath, with the aim of improving fermentation quality, reducing the quantity of effluent draining from the silo, and producing a fodder which stock will eat in large quantity if desired, so increasing potential production of milk or meat from the forage.

Disadvantages of wilting include more difficulty in choosing favourable weather for the work; greater crop losses in the field; more difficulty in avoiding high temperatures in the silo; and often a higher capital cost. Field losses are mainly mechanical; and with the use of improved field equipment there is no reason why any slightly higher field losses should not be balanced by the lower in-silo losses with wilted silage. Careful experiments at experimental husbandry farms have shown that on balance, wilting to a dry matter content of 25–30 per cent is worth while; but farmers who obtain good results with a simple direct-cut system find it difficult to believe that there can be much advantage in changing to a more complicated system. The weight of crop to be transported and ensiled is reduced by wilting, and this will normally result in fewer loads, and appreciable saving in transport and ensiling time. (Wilting from 20 per cent D.M. to 30 per cent reduces the weight to be handled by about one third.) On balance, wilting will usually require little or no more time and labour where suitable equipment is used. The main difficulty in switching to wilting is therefore the more difficult management caused by unsettled weather.

Wilted Silage Systems

Most of the systems discussed in connection with direct-cut silage making need little modification to adapt them for making wilted silage. In most cases it is simply necessary to allow for a cutting operation, 6–30 hours before picking up. The preliminary cutting may be done by mower, mower and crimper or crusher, flail forage harvester with 2-way chute, flail mower or rotary mower. Of these, the flail mower may be preferable where a simple flail forage harvester is to be used for picking up. The effect of any chopping and laceration during cutting is generally beneficial; and there should be very little loss of crop when collecting. However, the high speed of cutting by rotary mower is an important advantage.

Two-man Wilting Systems

Where two men are available full-time, the buckraking tractor may have time to do the necessary cutting; and this then makes a well-balanced system if a high cutting rate is achieved by using a fairly wide mower.

Three-man Wilting Systems

A good three-man system involves: one man with flail mower cutting; one man with in-line or side-mounted forage harvester running direct to the silo; and one man buckraking. The man who does the cutting will be available to use a second forage harvester for collecting when he has cut enough for the next day. Such a gang should be capable of an output equal to that of a direct-cut 3-man system where one man cuts and loads, one transports and one ensiles.

Four-man Wilting Systems

There is a good case for a four-man system where wilted silage is made. The team is made up of one man cutting; two with in-line or side-mounted harvesters and tipping trailers, each taking his own loads direct to the silo; and one man buckraking. This is a well-balanced team, capable of a working rate in the region of 2 acres (0·8 ha) per hour overall, (equivalent to 12–16 tons per hour of fresh material) and ensiling 70 to 100 tons of wilted material daily.

It should be added that very good wilted silage can also be made at a high rate in horizontal silos, using a full-chop forage harvester, the chopped material being tipped and buckraked on to the silo in the usual way. It is usually best to use a silo with tall airtight walls, and to seal with a plastic or elastic sheet well tucked down.

Making High-Dry-Matter Silage

Thirty per cent dry matter is about the upper practical limit for wilted silage made in a conventional type of horizontal silo, where there is no attempt at a positive air seal but steps are taken to minimize access of air by using solid silo walls and a substantially airtight sheet to cover the top. In such circumstances a considerable amount of consolidation by tractor rolling during making is usually needed, unless a vacuum compression method is used. At higher dry-matter contents it becomes impossible to control temperature by rolling unless a more positive sealing is achieved. Methods of sealing effectively include the use of a heavy sheet of an impervious material such as butyl rubber, in conjunction with a sound method of sealing between the sheet and the silo walls; or the use of a tower silo.

Where solid silo walls and an air-impermeable sheet are used, there is much to be said for rapid progressive filling in wedges from an end with a solid wall. This permits more effective exclusion of air during the vital early fermentation period.

The great advantage of a tower silo system is that it permits effective mechanization of silo filling, and exclusion of air from the crop during ensiling and storage. It also provides the basic necessities for efficient mechanized feeding, with sufficient silage removed daily in even layers, so avoiding unnecessary fermentation losses between storage and feeding.

Where a farm already has a reasonably efficient method of conservation based on a good self-feed silage system or a system of feeding barn-dried hay, the capital cost of scrapping the existing method in favour of a tower system is so high as to deter most farmers. This does not necessarily mean that a switch to a high-cost system can hardly ever be justified; but where capital is needed for developing production in other directions, careful calculations concerning costs and likely benefits are clearly necessary. The main factors which enter into such calculations are briefly discussed below; but there can never be a hard and fast conclusion which provides a universal answer to the problems involved in deciding on the most economic system. The more highly mechanized tower silo system inevitably incurs heavy machinery costs even for a small enterprise; but it becomes more economic with increasing size, up to the point where a single set of field equipment is fully employed. For such reasons there is no point in considering an installation smaller than that needed to cater for a herd of 60–70 cows, unless there is some arrangement for sharing use of most or all of the mobile equipment.

Advantages of the tower system include more efficient conservation; easy rationing if a suitable mechanized feeding system is incorporated; saving of waste in feeding; and saving of labour.

Experimental work in the U.S.A. and on Experimental Husbandry Farms in Britain has shown that there are considerable differences in efficiency of conservation between various types of silos. The extent of losses naturally depends on management, and so can never be a precise amount which is positively predictable. With this reservation Table 23 shows typical losses of dry matter for various types of silo. As a

Table 23. Typical Losses in Silage Making (Dry Matter, Per Cent)

Type of silo	Method	Loss in field %	Effluent %	Losses in Silo Ferment- ation %	Surface Spoilage %	Total in silo %	Total Loss %
Tower	High D.M. (35–40%)	8	0	5	0	5	13
Well-sealed clamp	Wilted (25–30% D.M.)	5	0	10	5	15	20
Well-sealed clamp	Direct (18–20% D.M.)	3	2	13	5	20	23
Vacuum clamp kept sealed	Direct (18–20% D.M.)	3	2	13	2	17	20
Unsealed clamp or stack	Direct (18–20% D.M.)	3	2	13	15	30	33

general rule, losses of starch equivalent will be substantially higher. With care, losses in a sealed tower silo can be kept round about 5 per cent D.M. There is an appreciable field loss associated with wilting to 35–40 per cent D.M. and then picking up by full-chop forage harvester. This can run as high as 20 per cent, but should normally range from 6 to 10 per cent, giving a total dry-matter loss of the order of 13 per cent. By comparison, the best practice with a horizontal silo in which wilted silage is made and self-fed results in a total loss in the region of 16 to 20 per cent, while the average of reasonably good practice results in about 25 per cent loss.

In practice, a switch to a tower silo system can result in advantages greater than the mere comparison of losses suggests. This can be due in part to the greater efficiency of feeding, and especially avoidance of waste. It can reasonably be assumed that a 10 per cent improvement in the amount of dry matter available for feeding can be expected from a well-managed tower system. The financial effect of this saving can be assessed in terms of the number of cattle that can be fed in winter, or the area of land which can be devoted to cash crops if the herd size is kept constant.

The saving of labour achieved by a mechanized feeding system depends on a wide range of management factors, as well as on machinery performances. Introduction of a mechanized feeding system is not usually aimed mainly at reduction of labour cost, though this may be a factor in a large-scale enterprise. On a family farm a major aim may simply be to reduce the amount of time spent on the job. As a general rule, the substitution of push-button methods for manual work will result in an increased number of stock handled per man, and in a welcome reduction of physical labour.

The potential labour-saving advantages of mechanized methods may fail to materialize if there are serious mistakes in planning of layouts, or if machines continually break down. Mechanized feeding systems often take some time to reach their full potential, because of deficiencies in the machines in relation to the conditions of use, or because of lack of knowledge or care on the part of the user. In calculating saving of labour, it has to be realized that completely automatic feeding of forage, though clearly attainable, is not achieved with most modern equipment. For example, though a silo unloader can be automatically controlled, most unloaders need occasional manual adjustment of the setting. It is not practicable to ration accurately on a simple time basis. Yet with good equipment on a well-run farm, it is possible for the operator to keep an eye on the mechanical feeding and to attend to several other routine jobs while the feeding is being carried out.

On some farms, feeding by fixed mechanical conveyors may not be practicable, especially where beef cattle are housed in a number of widely scattered yards. In such circumstances, where feeding by forage box is necessary, the saving of labour is less, but equipment cost is relatively low.

Disadvantages of a tower silo system include extra cost for machinery used to fill and empty the silo, and to convey food to the stock. The extra capital cost of a typical tower installation may be about £5000 more than for a self-feed system, resulting in an increased annual charge of about £825 per annum.

This is a substantial sum, but one which can be justified if the equipment is effectively used. On many farms, introduction of a tower system has resulted in an increase in the number of cows kept considerably greater than that corresponding to the savings in silage losses outlined earlier. In practice, efficiency of forage use tends to be higher than calculated, and there is little difficulty in making the same acreage conserved suffice for up to 20 per cent more cattle.

Grass Drying

The economics of grass drying in Britain can be argued from many standpoints, since drying may be done by farm or factory plants, while the dried grass may be either used on the farm or sold. Most of the 100 000 tons or so of dried grass produced annually in Britain is sold for other purposes than feeding of cattle. Much of it goes into rations for poultry; and for this and similar purposes, the carotene content is of high importance. There can be no doubt that dried grass is an excellent food for high-producing cattle; but it is difficult to justify the high cost of producing it for this purpose except as a cash crop when almost as good results can be obtained more cheaply by processes such as barn hay drying. When considering grass drying for the production of cattle food, any study of the economics of the process must pay due regard to the fact that the product may contain over 20 per cent protein and be worth over £60 a ton, or may contain less than 14 per cent protein and be worth as much as good hay. Estimates of the cost of drying can also vary widely, according to the method used for calculating depreciation, the annual use and efficiency of the plant, and the technique of management.

The target must usually be a yield of 4–5 tons per acre (10–12·5 ton/ha) dried grass with a minimum protein content of about 16 per cent, and with a high proportion of the carotene preserved. The value of such material will be around £60 a ton, and the cost of drying is discussed below.

An efficient modern drier is expensive, and can only justify itself if it is fairly fully used. 'Full' use in the case of a grass drier means something like 2500 hours annually; and it would be inadvisable to buy a new plant with the object of using it for less than 1000 hours a year. A $\frac{1}{2}$-ton per hour farm-scale plant is likely to cost about £25 000 to install. 1000 hours of use should produce 500 tons of dried grass, and assuming say a 10-year life at this low annual usage, annual depreciation will be £2500, or £5 a ton. With interest on capital at £2 a ton added, fixed charges are £7 a ton. The investment is, of course, a much better proposition if the drier is used twice as long, or is also used for grain.

Typical production costs for a large-scale co-operative grass drier, with capacity of $1\frac{1}{2}$–2 tons per hour of dried cobs, annual output of 4000–5000 tons dried product, and capital cost of equipment for harvesting, transport and drying of £125 000–£175 000, are as follows:

Cost of growing the crop	£15 per ton dried
Cost of drying and processing (including harvest and transport)	£25–£30 per ton
Total production cost	£40–£45 per ton

Typical breakdown of drying and processing costs for a large-scale co-operative drier, before the increases in fuel costs in 1973–4, were:

Item	% total harvesting, transport, drying and processing costs
Fuel	26
Labour and management	20
Depreciation	20
Maintenance and repairs	20
Electricity	6
Office and insurance	6
Bank interest	2
	100

Fuel represents the highest single cost in grass drying. If electricity is excluded from consideration as it must be on grounds of cost, the easiest fuel to use in a small farm plant is gas oil; but this is usually considerably more expensive than heavy fuel oil which is normally used. For the larger installations coal is a possible fuel, but its use is generally confined to very large factory-scale plants where it is economic to install automatic stoking equipment.

The prices of both oil and solid fuels vary according to the distance from the main distribution centres, and with coal the differences between one region and another can be over £2 a ton. Heavy fuel oil requires a more expensive furnace than gas oil, and it is necessary to provide a heater to keep the oil fluid. A large storage tank is needed.

The importance of reducing the moisture content by 'wilting' must now be studied, since by this means some types of low-temperature farm driers can be operated in such a way that only about 60 gallons (250 l) of fuel are needed to produce 1 ton of dried grass. The effects of wilting are clearly brought out in Table 24, which shows that the output of a drier is approximately doubled by reducing moisture content in the field from 85 to 75 per cent. 85 per cent moisture is fairly common in young crops collected in a wet condition, and even a reduction to 80 per cent brings about a considerable reduction in drying cost. The moisture is a little more difficult to remove from

Table 24. Effect of Field Wilting on Output of a Farm Grass Drier (Tray or Conveyor Types)

Moisture Content Wet Material	Water Ratio (to 10 per cent Moisture)	Dried Product at 10% M.C.	
		1 ton wet produces	Output of drier evaporating 1 ton water/hr
		tons	tons/hr
90	8·0	0·11	0·12
85	5·0	0·16	0·20
80	3·5	0·22	0·28
75	2·6	0·28	0·39
70	2·0	0·33	0·50
65	1·6	0·38	0·62
60	1·2	0·45	0·80
50	0·8	0·55	1·25

wilted grass, but so far as drying stock food for home consumption is concerned, the slight loss of carotene is insignificant and there can be no doubt that the practice is worth while for small farm-scale plants when weather conditions permit.

It is, of course, essential to be able to wilt the crop evenly, and this involves using modern swath treatment machines.

With some types of driers there is the possibility of using the plant as a barn hay drier in really favourable weather conditions. For example, a conveyor drier with a normal output of $\frac{1}{2}$ ton per hour at 80 per cent moisture (a normal grass drying figure) will produce about 2 tons per hour if the crop is wilted to 50 per cent moisture content (see Table 24). This is a reasonable way to use a farm plant for a week or two in the year while weather conditions are particularly favourable and crops are tending to run to seed. Unfortunately, most high-temperature driers are not suitable for drying wilted crops.

Barn Hay Drying

Barn hay drying consists essentially of bringing partly dried hay under cover and completing the drying process in conditions where rain cannot interfere, and mechanical losses of the valuable leafy parts of the crop are largely prevented. The process is essentially different from grass drying in that most of the moisture is removed from the crop in the field, and only cold or slightly warmed air is used to complete the drying.

Barn drying, intelligently applied, can remove a great deal of the risk from haymaking, and at the same time yield a product of a quality appreciably better than that of good field-made hay. The mechanical losses due to leaf shattering in field-made hay can be severe. Lucerne hay made in the conventional way loses 20–30 per cent of its leaf during making and a further 5–10 per cent during feeding. Moreover, the shattered leaves contain approximately twice as much protein and half as much fibre as the hay that is ultimately fed, which often consists largely of stalks. Leaf shattering is also apt to be serious in all types of clovers, and is not by any means negligible in grasses. In the barn drying process the crop is collected before shattering becomes serious, and this is one of its important advantages.

Broadly speaking, barn drying methods can be divided into (1) storage drying, in which the crop is gradually loaded into the barn and is left there until required for feeding and (2) a batch system in which drying is more rapid. As a general rule, cold air only is used for storage drying, while warmed air may be used for the batch system.

The need to carry out field work effectively is clearly indicated by the figures in Table 25, showing the amounts of water that still have to be removed from the crop if it is brought into the barn at various moisture levels.

Table 25. Water Removed by Barn Drying of Hay

Moisture Content of hay %	Weight of Water to be removed to produce 1 ton dried hay at 20% M.C.	
	cwt	kg
60	20	1010
55	16	810
50	12	610
45	9	460
40	7	350
35	5	250
30	3	150
25	1	50

It is obviously desirable to get the moisture content down to about 45 per cent in the field, otherwise drying time in the barn is bound to be prolonged, and the cost must be heavy, especially in adverse weather conditions. While batch driers with a high air flow and a heater can, if necessary, deal with moisture levels as high as 55–60 per cent, this is usually uneconomic. With storage driers it is difficult to deal with crops taken in at over 50 per cent m.c. Drying time in the barn naturally depends on batch size, air flow and temperature rise employed. Developments in field haymaking techniques have greatly increased the likelihood of successful utilization of barn hay driers, since so much depends on rapid and effective removal of moisture in the swath.

Table 26 shows the main characteristics of typical walled barn driers for baled hay. Such driers have a plenum chamber at the bottom, and airtight walls, so that the drying air passes

Table 26. Characteristics of Barn Driers for Baled Hay
       (Vertical Air Flow)

Storage Type		
Height of airtight walls	20 ft	6 m
Area of bay	400–900 ft²	35–85 m²
Fan performance	45 c.f.m./ft²	0·23 m³/m²/s
at resistance	3 in w.g.	7·5 mbar
Air temperature rise		
Normal	nil	nil
Maximum	10 °F	6 °C

Walled batch Type		
Height of air-tight walls	10 ft	3 m
Area of bay	200–450 ft²	18–40 m²
Fan performance	60 c.f.m./ft²	0·3 m³/m²/s
at resistance	3 in w.g.	7·5 mbar
Air temperature rise		
(max., often at reduced air flow)	25 °F	14 °C

more or less vertically from bottom to top. In another type of
barn drier, described later as the Dutch type, air is introduced
into a chimney built up the centre of the stack of bales; no side
walls are used, apart from well-spaced supports for the stack of
bales; and the air flows more or less horizontally from the
centre outwards. In the centre-duct system the air flows in a
fan-shaped pattern from a duct running above-ground along the
centre of the building. As a general rule it is essential for both
the Dutch and the centre-duct system to reduce loading moisture content to at most 40 per cent.

General Management – Vertical Airflow System

Effective barn drying depends on producing bales of a uniform
moderate density, of the order of 10–12 lb per cubic ft (160–
210 kg/m³), at not over 45 per cent moisture content, and
preferably less. In good drying conditions the aim should be
not over 35 per cent m.c. For full-length bales typical figures
should be:

Bale length	36 in	0·9 m
Bale section	18 by 14 in	0·45 by 0·35 m
Bale volume	5¼ ft³	0·15 m³
Wet weight	52–63 lb	24–29 kg
Dry weight	36–48 lb	16–22 kg

Bales handle better after drying if they are made only about 30 in (0·76 m) long.

When loading the drier it is advisable to start blowing as soon as the floor is covered, and to aim to get the bottom reasonably dry before completing filling the bay. Thus, in a multi-bay plant it may be best to start filling a second and even a third bay before the first is completely filled. Air flow falls as loading depth increases, and there is a danger, if very damp bales are loaded too deep, that the compression will create a very high resistance which makes it impossible to force air through. If this happens, hay can soon be spoiled. A light handkerchief spread over the top of the hay should be slightly raised by an air flow of 45 c.f.m. per sq ft (0·23 m³/m²/sec). A simple water manometer connected to the plenum chamber provides a handy guide to loading. With most fans, resistance should not reach 3 in w.g. (7·5 mbar).

Completion of drying is best ascertained by switching off the fan overnight when it is thought that drying is nearly complete, and observing whether there has been any heating by standing on top of the hay when the fan is switched on next day. This procedure may be repeated until it is clear that there is no spontaneous heating. There is a common tendency on the part of those new to barn drying to expect that the hay is dry long before drying is complete. With a highly-nutritious early-cut hay, however, the whole has to be reduced to below 20 per cent moisture content; and in adverse conditions this may take several weeks.

'Dutch' (horizontal air-flow) Hay Drying System

In the 'Dutch' system the air is distributed horizontally through the crop, from a vertical flue formed in the stack of bales. The fans are usually suspended from the roof, but the air may be introduced from a below-floor tunnel, the top of the flue then being closed by a 'bung'. Typical design features are:

Maximum stack height	20 ft	6 m
Maximum stack area	900 ft²	84 m²
Air flow per unit of floor area	35 c.f.m./ft²	0·18 m³/m²/s
at resistance	2·5 in w.g.	6·2 mbar

Centre Duct Drying System

The centre duct system is designed for use in open-sided buildings to dry hay wilted to not more than 40 per cent moisture content. A framed duct with open sides and top runs centrally down the length of the barn. Bales are stacked over this duct, tunnel-fashion, and cold air is blown through at 40 c.f.m. per sq foot of ventilated floor area. The height and width of the air duct must match the barn dimensions so as to keep the lengths of the various air paths as equal as possible. Thus, a suitable duct section for a stack 30 ft wide and 20 ft high is about 3 ft wide and 6 ft high. Bales should be stacked round the duct over the full length of the tunnel so that drying can begin as soon as possible. Typical design features may be summarized as follows:

Maximum stack height	20 ft	6 m
Maximum stack area	900 ft^2	84 m^2
Air flow per unit of floor area	40 c.f.m./ft^2	0·2 m^3/m^2/s
at resistance	3 in w.g.	7·5 mbar

Economics of Barn Hay Drying

The capital cost of providing equipment for barn drying is very variable. At one extreme, a farmer who has a suitable walled barn, and builds himself a plenum chamber from secondhand materials, can do the whole job for little more than the cost of providing a fan and connecting it to the electricity supply. Alternatively, he may buy a tractor fan and use a spare tractor, provided the latter has a good engine and p.t.o. At the other extreme, it may be necessary to provide a new barn as well as the complete drying installation. A 60-cow herd will need a 3-bay barn, of overall dimensions about 45 ft by 30 ft by 20 ft high, which, if well filled, will hold about 90 tons of hay. The needs of a slightly larger herd may be catered for by a barn 60 ft (18·3 m) long by 30 ft (9·2 m) wide and 21 ft (6·4 m) to eaves, with a capacity of about 120 tons. The following figures show estimated approximate capital costs of typical installations based on buildings 60 ft long and 30 ft wide.

Type of drier	In new building, 21 ft (6·4 m) to eaves, capacity 120 tons. Cost, £	In existing building, 18 ft (5·5 m) to eaves, capacity 100 tons. Cost, £
Mesh floor, 2-bay, single fan, 40 h.p.	3350	1850
Mesh floor, 2-bay, two fans, 40 h.p. each	3950	2450
Dutch system, 2 bays, single fan, 30 h.p.	2600	1100
Centre-duct system, single 50 h.p. fan	2700	1200

In making comparisons, account should be taken of the differences in possible handling methods and in maximum advisable moisture content at loading already mentioned.

If the capital cost of a new 120-ton installation is £3000, depreciation on the basis of a 10-year life is £2·50 per ton and interest charge £1 per ton, giving a total of £3·50. Addition of £2 per ton for running cost gives a total cost of drying and storing, excluding labour and field machinery, of the order of £5·50 per ton. These figures indicate the advantage, where it is practicable, of making use of low-cost buildings. Where an existing barn can effectively be used in conjunction with the Dutch or centre-duct system, capital cost can be of the order of £10 per ton capacity, and on this basis annual cost with a 10-year life can be below £2 per ton. So with a running cost of about £2 per ton, total drying and storage costs around £4 per ton can be achieved. This is a small amount to pay for the advantages that barn drying offers to a dairy farmer who relies on his own hay for winter feed.

Use of Heat in Barn Drying
Some barn driers have a heater, which may be electric, indirect oil-fired or may utilize the waste heat of the engine which drives the fan. As a general rule, cheapness in drying is achieved by using large amounts of atmospheric air, rather than by warming up a small air flow. It is therefore important to site barn driers away from misty valleys, where the air is often much damper than the average of the district.

It is advisable to resist the temptation to save cost by making use of spontaneous heating of the hay to assist drying. This uses up the valuable carbohydrates in the hay, and calculations show that the heat so provided is more expensive than that obtained by using an electric heater. Thus, in unfavourable weather it is advisable to do sufficient blowing to keep the hay cool. In such conditions, it is useful to be able to warm the air a little; but if the source of heat is electricity it is generally necessary to use heat sparingly in order to avoid excessive drying costs.

In general, drying with cold air is cheaper than using electrical heating save in extremely cold and damp conditions; but the use of electric heating can be economic in the final stages of drying in a batch drier, especially when the weather is cold and wet. Heating the air is, of course, likely to be more often economic where a cheap source of heat is used.

The waste heat method is attractive in view of the ever-increasing number of 'spare' tractors that are available on farms. A tractor with a vaporizing oil engine of fairly low thermal efficiency is ideal for the job, provided that the engine is in good order. Diesel engines, having a higher thermal efficiency, give a somewhat lower temperature rise in corresponding conditions. The extent of the temperature rise that can be secured should not be exaggerated. Taking the case of a reasonably efficient diesel engine, the temperature rise may be reckoned as follows:

Diesel fuel has a net calorific value of about 150 000 b.t.u. per gallon or 19 000 b.t.u. per lb (4·4 MJ/kg). An engine which uses $\frac{1}{2}$ lb of fuel per belt h.p. hour uses fuel of 9500 b.t.u. calorific value per h.p. developed, and of this, 3 100 b.t.u. produces useful work, about 2300 b.t.u. goes into the cooling water, 2300 goes out with the exhaust gases and the balance of 1800 b.t.u. is dissipated by way of conduction and radiation. The total of 'waste heat' is 6400 b.t.u. per hour, and of this it is possible to collect about 75 per cent or 4800 b.t.u. per h.p. (1·5 MJ/kW) developed.

Now suppose we have a tractor developing 20 h.p. in driving a relatively inefficient centrifugal fan which delivers 20 000 c.f.m. against 2 in w.g. when drying baled hay. Heat available = 4800 × 20 = 96 000 b.t.u. per hour. Now 1000 c.f.m. of air needs just over 1000 b.t.u. to maintain a temperature rise of 1 °F for 1 hour. Therefore 20 000 c.f.m. will need just over 20 000 b.t.u. per hour to maintain a 1 °F temperature rise, and the temperature rise provided by 96 000 b.t.u. will be 4–5 °F. Owing to the inefficiency of the fan employed, there will be a temperature rise of 1–2 °F provided by the fan itself, so the total temperature rise can be 5–7 °F (3–4 °C) in these conditions. This is not large, but is adequate to ensure that in most conditions the air will do a satisfactory amount of drying.

Benefits from Barn Hay Drying

In ideal weather conditions, field-dried hay can be almost as good as, or occasionally even better than, the barn-dried product. On the other hand, in normal British weather conditions barn drying provides a reliability which is of great value to a dairy farmer who depends on hay for winter feed. One positive advantage which can be measured has been recorded in several long-term experiments on Experimental Husbandry Farms is the extra yield of nutrients that barn drying can help to produce. Table 27 shows the increased yield obtained at Drayton E.H.F.

Table 27. Seasonal Average Yields of Field-dried and Barn-dried Hay at Drayton Experimental Husbandry Farm (Warwickshire)

| | Dry Matter Yield | | | | Increased Yield from Barn Drying as % of Field Dried |
| | Field-dried hay | | Barn-dried hay | | |
Year	cwt per acre	kg/ha	cwt per acre	kg/ha	
1958	43·3	5400	51·2	6400	18
1959	41·4	5200	47·4	5900	14
1960	39·2	4900	43·4	5400	11
1961	28·1	3500	30·7	3800	7

Zero Grazing

Though this chapter is concerned primarily with fodder conservation, it is appropriate to mention here a further use of forage harvesters for the technique sometimes called 'Zero Grazing', in which the crops are cut and fed to the stock in yards in summer as well as in winter.

Extra costs can be roughly calculated on the basis of the cost of cutting and collecting crops for silage by using a forage harvester. Extra costs will probably be in the region of £0·50 per ton of green crop, and this with a production of say 8 tons during a season would mean some £4 per acre. There is also the extra cost and work involved in handling the manure from the yarded cattle. This will usually be in the form of slurry, and distribution costs have to be considered. There may be palatability problems if the slurry is applied too near to cutting time. Extra returns are problematical. Experiments in U.S.A. have produced variable results. In Rhode Island, only 1·12 acres (0·45 ha) were needed per cow compared with 1·37 (0·55 ha) under rotational grazing, and the increased yield amounted to 571 lb of milk per acre (640 kg/ha). In Minnesota results were considerably better. Only 0·87 acres (0·37 ha) were needed per cow compared with 1·45 acres (0·59 ha) under rotational grazing. Increased milk yield per acre in this case was 1013 lb.

The practice has been successfully applied on farms where stocking rates and yield are high and there is intensive use of nitrogen fertilizer. Zero grazing or storage feeding are probably the only techniques by which such intensive production can be effectively used by some farmers. Where the land is grazed, the high concentration of droppings can spoil a high proportion of the herbage, and the only practical alternative is a very intensive paddock grazing system.

A further argument in favour of zero grazing or storage feeding is related to the management of very large herds. With numbers about 150, physical control of grazing becomes difficult, and poaching of the land near field gates a serious problem. By comparison, cattle control in well arranged yards, with mechanical feeding, is relatively straightforward.

At High Mowthorpe E.H.F. it has been found that zero grazing can be successfully practised with young beef cattle for the five months May to September. When the cattle were kept in yards, slurry handling or provision of sufficient bedding caused serious problems; but it was later established that the stock could be kept outdoors, on a small area of exposed chalk land. One and a half to 2 man hours are needed daily to look after about 120 cattle, which are normally kept on a small 'sacrifice' area of less than 2 acres (less than 0·8 ha) which is due to be ploughed from ley in the autumn. Preliminary trials showed that a given area of forage produced about 10 per cent more carcase weight when zero grazed than when grazed in the normal way. The only special equipment needed beyond that normally used for silage-making was a self-unloading forage box. An advantage of the system is the ease with which the stock take to silage feeding when fresh grass is no longer available.

Mechanical feeding of cattle in summer is most likely to be applied on farms where tower silos and mechanical feeding are used during the rest of the year. On such farms, if conservation is efficient, it may be advantageous to practice storage feeding throughout the year, rather than resort to zero grazing in summer. Storage feeding should permit harvesting forage crops at the right stage of growth, and it gets rid of the need to collect grass every day, wet or fine, throughout the summer. A disadvantage is that extra silo capacity is needed.

CHAPTER EIGHT

Harvesting Corn, Seed Crops and Vegetables

Grain Harvesting. Combine Harvester Capacity. Cost of Combine Harvesting. Grain Drying and Storage. Choice of Type of Drier. Estimation of Grain Drying and Storage Costs. Grain Storage at High Moisture Content. Harvesting Seed Crops. Harvesting Peas and Beans. Harvesting Green Peas for Canning or Freezing. Harvesting Cabbage and other Brassica Crops.

The aim of mechanization in the corn harvest, as in the hay harvest, should be to equip in such a manner that the regular staff can get on rapidly with the work when weather permits, so that crops can be gathered in good condition. Speed is important, since delays can soon result in serious losses owing to the crops being spoiled. The corn from an acre of land can be expected to be worth £100 or more (£250+ per hectare); so it is clear that even on medium-sized farms a considerable expenditure on equipment can be justified if its use will ensure safe harvesting at low cost.

Grain Harvesting

The switch from harvesting by binder and stationary thresher to harvesting by combine has become a matter of history. The first trial with a combine harvester in Britain took place in 1928, and by the end of the 1930s it was certain that the method had come to stay. It was not, however, until the 1950s that combining began to affect corn growers of all sorts and sizes, and it became clear that the binder method would inevitably disappear altogether. The main reason was saving of labour. Studies carried out when the binder method was still important in Britain showed a total labour requirement for harvesting, stacking, thatching and threshing in the region of 25 man hours per acre (72 man hours/ha). Comparable figures for a modern well-mechanized system total only about 20 per cent of this, including straw handling. Modern problems are concerned with doing the work in good time without any extra staff. Typical questions are:

How much combine capacity is required for a given amount of crop, and what will this cost?

Can the capital investment be justified, or are there practicable ways of minimizing investment, e.g. by using older machines, or by joining a machinery syndicate?

Would it be more economic to rely on contract work?

What are the costs and likely returns from long term storage of grain, and what storage methods are best suited to the individual farm's needs?

Combine Harvester Capacity

Combine harvester capacity is so variable, according to crop conditions, that manufacturers' unsupported claims must always be subject to some reservation. The best sources of reliable information are tests carried out by official testing stations, working to a standardized testing procedure, where one of the major aims is to produce results which can be compared. Even with such tests it is necessary to study the reports with care, taking note of any exceptional circumstances. Always, in studying the reports of tests, care must be taken to distinguish between what are called 'net' (or spot) outputs, achieved in rating runs, and the much lower 'overall' outputs.

The rating runs, usually carried out at about 50 lb and 100 lb per acre (56 and 112 kg/ha) grain loss, indicate a limit to the output when there is nothing to prevent continuous work. The overall working rate which a farmer can expect to achieve is generally only about half the rating figure.

In comparing reports of tests carried out in different seasons, it should be remembered that badly laid and externally damp crops will always slow down the rate of work, and there is at present no formula by which it is practicable to 'correct' results for crop conditions. A high grain moisture content does not slow down the combine unless it is associated with wet straw or laid crops.

It can be misleading to relate output of a combine to cutting width; but with typical machines, there is a range from about 2–3 tons per hour from a 6 ft (1·8 m)-cut p.t.o.-driven trailed machine, to 10–15 tons per hour for very high output machines of 12–30 ft (3·5–4·5 m) cut, with very powerful engines, e.g. over 100 h.p. Most modern machines are self-propelled and fall into a group having 8–12 ft (2·5–3·5 m) cut, an engine of 50–60 h.p. and an output of 5–10 tons per hour. Since the prices of these machines range from about £3000 to over £10 000, it is obvious that farmers need to make a fairly careful study of performance in relation to price, with the aim of securing a reasonable balance between ability to harvest crops in good time, and avoidance of wasteful capital expenditure.

When every effort has been made to relate combine capacity to the need, the farmer has to make a decision that takes account of many factors which simply cannot be assessed in arithmetical terms. For example, the number of hours during which the grain is fit to harvest in a wet year may be less than half the time available in a particularly fine harvest. Most farmers cannot justify equipping to avoid every worry in the worst harvest, and in practice the decision made is nearly always a compromise, which is pushed towards more insurance or more risk according to individual needs.

Even the number of combining hours 'available' in given weather conditions is not a firm figure which is the same for all farms. Practical considerations which affect it include the drying arrangements on the farm, and the willingness or otherwise of operators to work at all hours of the day or night, or at weekends. The number of combining hours reasonably available in the East Midlands has been estimated at 175–200 in an average season; but many farmers in such a season only wish to work about 150 of these. Even so, a fairly small machine, capable of $1\frac{1}{4}$ acres (0·5 ha) per hour overall, can manage almost 200 acres (80 ha) per season, while a machine with an average capacity of 6 tons per hour can handle around 500 acres (200 ha) in most conditions. The penalty for failure to achieve timely harvesting can sometimes be severe. In a trial on an Experimental Husbandry Farm, yield of a crop of barley fell by 4 cwt/acre (500 kg/ha) in a week of good August weather during which there was practically no wind or rain.

Cost of Combine Harvesting
One method of calculating the cost of combining is discussed and illustrated in Chapter 3 (Table 9), where estimates are made of operating cost at various seasonal outputs of a high-capacity machine costing £6000, a small, new machine costing £4000, and a secondhand machine bought for £2000. The figures illustrate how heavily fixed costs of the more expensive machines weigh at low annual usage, and show that higher repair costs for a moderately priced secondhand machine can be relatively unimportant. It should, however, be remembered that the secondhand machine can hardly be expected to go right through the season without giving a little trouble, even if it has been well overhauled before the season begins.

The trade in combine harvesters is in many respects like that in cars, commercial vehicles and tractors. The machine's value falls sharply as soon as it starts work; and its secondhand value in its early years is related almost as much to the year of manufacture as to the amount of work done. A special feature of the combine market is the very short season of use, and the out-of-season discounts offered by manufacturers in order to ease distribution problems. In these circumstances it can occasionally be useful to calculate ownership and operating costs on the basis of true depreciation, rather than on the simpler figures provided by Tables A.2(i) and A.2(ii). In a study carried out by Leeds University it was concluded that true depreciation of combines was very close to figures obtained by calculating an annual fall in value of $22\frac{1}{2}$ per cent, based on the diminishing values. Since this study was made, the market for both new and secondhand machines has become more 'saturated'; and there is a tendency for older machines to become relatively less valuable, as larger farmers change to a smaller number of higher-capacity machines. In these circumstances, 25 per cent of the diminishing value is probably nearer present day facts, and this is an easier figure to use. On this basis, the value of the secondhand machine would fall as follows:

	Capital value £	Annual depreciation £
New price	4000	—
Value after 1st harvest	3000	1000
Value after 2nd harvest	2250	750
Value after 3rd harvest	1688	562
Value after 4th harvest	1266	422

Most combine harvesters do not approach the potential 200 or so hours of use. Typical annual use on farms where cereal-growing is an important enterprise has been found by A.D.A.S. investigations to be 100–150 hours per season: but the average for all combine harvesters is less than 100 hours per season. In fact, the total amount of work to be done on many small farms is well below 100 acres (40 ha), and the advantages of buying a little-used and well-cared-for machine after it has done a few harvests are obvious.

In calculating the repair costs of secondhand machines, the figures suggested in Table A.3 may be used as a guide, but it must be remembered that these percentage figures should be applied to the NEW price. There will, of course, be a tendency for repair costs to increase with age, as the depreciation falls; but up to the end of the fourth harvest, there is no reason to expect repair costs greatly to exceed those calculated on the basis of Table A.3, unless there has been unusual neglect, accident or mis-use.

Machinery Syndicates for Combine Harvesting
It might be imagined, and is indeed often asserted, that a machine for use in a very limited season, such as a combine harvester, would be unsuited to syndicate use. In practice, however, combine harvester syndicates have almost invariably been thoroughly successful. It is, of course, necessary to adhere to the principle of not attempting too large an acreage in relation to combine strength. If this is done, the syndicate offers the possible attractions of limited capital investment; use of a high-capacity labour-saving machine; and, by labour substitution, use of a skilled operator who has a reasonable chance to learn how to use the machine efficiently, and so avoid

unnecessary waste of time and of grain. Many small-scale users just begin to re-learn how to use their machines by the time harvest is finished. The potential cost advantages of syndication to a small-scale farmer are well indicated by the figures in Table 9, where cost per acre at various levels of annual usage may be compared. Cost per acre falls steeply with increasing usage up to about half the potential capacity. Beyond this point, the main advantage of increased annual use is some further saving of capital investment.

Combining by Contract

Combine harvesting can be well suited to contract work where the contractor is conscientious in such matters as avoiding attempting too large an acreage, and so having to work when crop conditions are unsuitable. From the farmer's viewpoint, the work must be done in such a manner as to avoid losses, and the price needs to be reasonable in relation to the area to be worked. When contract work is compared, from a farmer's viewpoint, with a syndicate, it has to be recognized that the contractor is in business for profit; and assuming that the contract service and syndicate are equally well run, there is likely to be a smaller profit margin from farming after the services of an independent contractor have been paid for. From the viewpoint of satisfying the needs of a group of farmers, there is no basic difference between contract work and a syndicate; but in practice syndicates, with their guide rules concerning an equitable rota to fall back on if necessary, seem on the whole to result in less worry and grumbles than contract work. In spite of the difficulty that even a good contractor has in satisfying a number of independent clients, such services are important in some areas, and seem likely to remain so until farmers in general are keener on taking part in syndicates.

Grain Drying and Storage

With development of the use of combine harvesters it has become necessary to provide storage for many millions of tons of grain, most of which needs some drying or conditioning before it can be safely left in a bulk store. Some countries, faced with similar problems, have built a relatively small number of large stores, which may be operated by the grain marketing authority or by the State; but early experience in Britain of nationally operated silos underlined the size of the drying problem in a large installation in an adverse season; so though there are many large driers operated by merchants and a few by co-operative groups, the main trend in Britain since 1945 has been towards drying and storage on the farm. Table 28 shows how the population of farm driers has steadily risen, as combine harvester capacity increased.

Table 28. Numbers of Combine Harvesters, Pick-up Balers and Grain Driers on Farms in England and Wales

	1956	1962	1964	1965	1970	1972
Combine harvesters	31 030	52 350	—	57 950	56 670	49 990
Pick-up balers	37 810	76 940	—	87 410	85 160	78 540
Grain driers, total	7 690	19 710	27 010	—	—	—
continuous flow ⎱	2 550	7 260	9 790	—	11 770	—
tray or batch ⎰	—	2 510	3 710	—	7 340	—
platform (in sack)	2 320	4 190	3 680	—	—	—
ventilated bin	— ⎱	— ⎱	— ⎱	—	14 920[a]	—
floor driers	2 820 ⎰	5 750* ⎰	9 830* ⎰	—	10 260	—

* Questions asked but answers considered unreliable.
[a] Figure represents number of farms using type.

The moisture content required for storage depends on several factors, such as whether the corn is stored in sacks or in bulk, and on the amount of mechanical damage done to the grain in the harvesting operation. For general practical purposes the figures in Table 29 may be used as a guide.

Table 29. Maximum Grain Moisture Content Recommended for Various Storage Conditions at Normal Temperature (60 °F) (13 °C)

	Percentage Moisture Content			
	Seed corn and Malting barley		Other grain	
Duration of storage	Unventilated	Well ventilated	Unventilated	Well ventilated
Up to 4 weeks from harvest	16	17	17	18
Until February	14	16	15	17
Until April	14	15	15	16
Beyond April	13	14	14	15

Table 30. Types of Grain Drier

Main Type	Chief Variations
1. *Continuous grain flow* (Grain layers shallow — usually not over 1 ft (0·3 m). Relatively high temperatures employed).	(a) Gravity-flow tower with vertical grain walls contained between parallel perforated sheets. (b) Gravity-flow tower with the grain contained between parallel perforated sheets arranged in a series of inclines. (c) Gravity-flow tower with air-tight walls and horizontal louvred inlet and exhaust ducts arranged alternately. (d) Gravity-flow incline with louvred bed. ('Cascade.') (e) Assisted gravity flow down inclined perforated tray, with chain and slat scraper. (f) Horizontal conveyor, including multiple-band type. (g) Horizontal or slightly inclined oscillating conveyor, with perforated or louvred bed. (h) Rotary drum.
2. *Batch driers for loose grain* (Grain layers usually 1–3 ft (0·3–0·9 m) thick. Temperatures not much above 100 °F (38 °C))	(a) Self-emptying, with parallel vertical perforated walls and inclined floor. (b) Self-emptying with hopper bottom and perforated or louvred inlet and exhaust ducts. (c) Self-emptying tipping tray with perforated bottom. (d) Inclined tray driers not completely self-emptying. (e) Flat tray driers not self-emptying. (f) Automatic self-emptying.
3. *Batch driers for grain in sacks* (Temperature rise usually about 25–30 °F (14–17 °C))	(a) Platform with individual sack apertures. (b) Continuous welded wire mesh or perforated floor on which sacks are placed close together. (c) Tunnel drier in which sacks form an enclosed air duct at right angles to the main air duct.
4. *Ventilated bins* (Grain layers usually 3 ft (0·9 m) or more thick. Air temperature usually not over 10 °F (6 °C) above atmospheric).	(a) Vertical air flow with air-permeable flat floor. (b) Vertical air flow with perforated hopper bottom. (c) Vertical air flow with air ducts resting on the silo floor. (d) Semi-vertical air flow with horizontal inlet and exhaust ducts in alternate layers. (Often self-emptying.) (e) Radial air flow with air-permeable centre duct and walls. (Can be self-emptying.) (f) Horizontal air flow with vertical inlet and exhaust ducts. (Can be self-emptying.)

Table 30. Types of Grain Drier—*continued*

Main Type	Chief Variations
5. *On-floor driers* (Grain depth usually up to 8 ft (2·5 m), with level top, and substantially vertical air flow). Air temperature not over 10 °F (6 °C) above atmospheric.	(*a*) Portable lateral ducts, usually spaced at 3–4 ft (75–100 cm) centres. (*b*) Other types in which air may be distributed from large central duct only, into heap of grain having substantially greater depth at centre than at sides.

The proportion of grain needing to be dried and the degree of drying that is necessary varies from district to district and from season to season. There is no part of Britain where it is safe to assume that drying will never be needed, but farmers in the South and East have considerably less drying to do than those in the North and West. A full account of drying problems, methods and equipment is given in a Ministry of Agriculture Bulletin.[17]

Classification of driers into only three or four groups as in Table 28, gives no idea of the vast range of types of machines and installations available. A more comprehensive classification is indicated in Table 30.

A study of this classification and of the machines themselves shows that there are no sharp dividing lines between some of the main classes. For example, there is no fundamental difference in construction between some batch driers and some ventilated bins. The differences between the groups lie mainly in the thickness of grain layer and the air temperature rise employed. With some plants these can be varied at will between wide limits. For example, it is possible with many ventilated bin plants to use one or two bins as tray driers when need arises.

Choice of Type of Drier

In view of the wide range in both machines and farming conditions, it is necessary to assess the suitability of the various main types of driers for different farming circumstances. On large farms the problem of choice is relatively simple. Bulk handling of grain from the combine is essential, so installations which dry grain in the sack need not be considered. Some forms of batch driers are a possibility, especially installations comprising one or more self-emptying drying compartments, with fully mechanized conveying. In the long run, a continuous-flow drier or an 'automatic' batch type is likely to be preferable in wet regions, where a considerable amount of moisture normally has to be removed, while in dry regions such as the Eastern Counties, personal preference can well decide the choice between such machines and an adequate ventilated bin or on-floor installation.

The choice between a continuous or batch drier and drying in bulk has to be made on many medium sized farms as well as on large ones, and it is worth while to consider the main factors which should be taken into account.

On many farms, grain is the most important commodity produced, and it is therefore necessary to ensure that the methods of harvesting, handling and storage are not merely effective, but are the most economic that can be devised. Even for the grain growing specialist, however, what is most economic for one farmer will not necessarily be so for his neighbour. Moreover, the solution chosen is often strongly influenced by personal preferences. Thus, one specialist corn grower may decide that because grain will always be his most important product, he is justified in equipping his farm in the best way technically possible to clean, dry, store, grade and weigh out the grain, all by press-button equipment. This can be very expensive in capital cost; and a neighbour who depends equally on grain sales may decide to go to the other extreme, and limit his activities to harvesting the crop, and selling it for what he can get, directly from the harvest field. Experience shows that both of these extremes are usually inadvisable, and that the most economic courses lie in equipping for drying and storage in an economical way. Fortunately, there is a choice of British-made equipment suited to doing the job in a great variety of ways, and in large or small quantities.

Surveys show that small and medium-sized farms are able to compete effectively with large ones in devising low-priced drying and storage installations. This is because they can often

Table 31. Examples of Capital Cost of Typical Grain Drying
and Storage Plants

Type of drier	Tons per year		Capital cost* £ per ton stored
	Dried	Stored	
Floor-ventilated indoor bins[1]	300	250	55
	600	500	45
	1200	1000	35
Floor-ventilated outdoor bins[2]	400	300	22
	600	500	20
Radially ventilated bins[3]	400	250	55
Continuous 6 ton/hr[4]	800	600	50
Continuous 6 ton/hr[5]	1000	800	35
On-floor, metal laterals[5]	200	200	38
	400	400	35
	800	800	30
On-floor, air emptying[6]	800	800	33
On-floor, multi-purpose, with below-floor laterals[7]	800	800	33

* Includes building costs, before deduction of grant, and installation.
[1] Rectangular, galvanized. Wet pit, cleaner, bucket elevators, chain and flight conveyors.
[2] Cylindrical, galvanized. Large-capacity mobile augers and sweep auger.
[3] Expanded metal. Wet pit, cleaner, bucket elevators, chain and flight conveyors. Capacity for rapid drying.
[4] Rectangular, galvanized bins. Wet pit, cleaner, bucket elevators, chain and flight conveyors.
[5] 8 ft deep floor storage. Mobile auger and thrower, for large plants.
[6] 8 ft deep storage; wet pit, conveyors and thrower.
[7] Close-spaced ducts suitable for onion drying and storage.

make good use of existing buildings, whereas large-scale farmers normally have to erect a new building. Moreover, the large-scale plants normally incorporate a considerable amount of fully mechanized handling equipment, and store most of the grain in bulk. Capital cost is influenced by amount and type of storage, and the complexity of the auxiliary equipment. The cheapest large-scale storage is provided by on-floor installations. On-floor storage in conjunction with a continuous drier tends to be a little more costly than on-floor drying and storage. Improvements in design and construction of equipment have tended to keep the cost of complete drying and storage installations fairly stable, while many other costs have risen rapidly. This has been partly due to the general substitution of prefabricated components for equipment which formerly had to be laboriously built by hand work on the farm.

Estimation of Grain Drying and Storage Costs

When an attempt is made to estimate what it costs to dry and store grain, account must be taken of labour requirements and fuel costs, but the result depends mainly on the capital cost of the installation and on what assumptions are made concerning calculation of the 'fixed' costs. There are two main methods of approaching the problem of fixed costs. By one method it is assumed that any capital invested in drying and storage must earn a high rate of profit, so that the capital invested can be written off quickly, and become available for re-investment in a short time, e.g. 5 years. For most farmers who have always grown corn and are likely to continue to do so indefinitely, this approach is considered desirable but unrealistic. Its use results in drying and storage costs which are considerably higher than those actually incurred on most farms. This approach is therefore not further discussed.

For most farms, the figures for useful life of grain driers given in Table A.2(i) provide a realistic basis. The amount of time spent annually on drying is variable according to the drying need in different seasons; but for the purpose of general calculations it is reasonable to assume that annual drying time averages about 200 hours, and that life of the equipment, for purposes of writing off capital and calculating annual depreciation, may be reckoned at 10 years. This is in accord with investigations reported by the Farm Economics Departments of Cambridge, Nottingham and Leeds Universities. This figure of a 10-year life can be applied to all of the machinery associated with grain drying, cleaning and handling.

Bulk storage bins and buildings require separate consideration. From the viewpoint of simplicity, there is much to be said for writing off the capital in 10 years, and not attempting to distinguish between machines and buildings when estimating depreciation. On the other hand, there may in particular circumstances be a good case for reckoning on a longer depreciation period for these items, e.g. 15 years. It is certain that if a farmer wishes to compare two drying and storage installations, one of which consists mainly of machines, while the other consists mainly of buildings, he can usually justify expecting a longer useful life from the latter. Nevertheless, the argument that even buildings should pay for themselves in 10 years is a valid one. If this is accepted, total annual fixed charges, including interest, amount to some 14 per cent of the capital cost. In the case of the plants listed in Table 31, annual fixed charges on this basis range from about £2 per ton stored for a fairly large drier to about £7 per ton for some of the more complex small installations. Table 32 shows typical annual fixed costs for

Table 32. Annual Fixed Costs of Grain Storage at Various Levels of Capital Cost. £ Per Ton

Capital cost per ton £	Depreciation Life					
	10 years			15 years		
	Depreciation	Interest	Total	Depreciation	Interest	Total
5	0·50	0·20	0·70	0·33	0·20	0·53
10	1·00	0·40	1·40	0·67	0·40	1·07
20	2·00	0·80	2·80	1·33	0·80	2·13
30	3·00	1·20	4·20	2·00	1·20	3·20
40	4·00	1·60	5·60	2·67	1·60	4·27
50	5·00	2·00	7·00	3·33	2·00	5·33

various levels of capital invested in storage, based on a 10- and 15-year life. The need to keep capital costs below the upper limits of this table is appreciated when the limited returns from storage are considered later.

As regards physical life of equipment, surveys have shown that some items, such as pneumatic conveyors, were very quickly superseded; and some types of building work can quickly become a liability when new machines are introduced. On the other hand, many machines over 15 years old give trouble-free service.

Running Costs

It has been seen that annual fixed costs are likely to be in the region of £3 to £6 per ton stored; and to these must be added the running costs. These may be almost negligible in a very dry season, but will be appreciable in a wet harvest.

Removal of 6 per cent moisture content from a ton of grain requires about $2\frac{1}{2}$ gallons (11 l) of diesel oil and 8 units of electricity with a reasonably efficient continuous flow or batch drier. With oil at 22p per gallon and electricity at 1·2p per unit, cost per ton is 65p. Cleaning and conveying may bring fuel and power cost to about 67p per ton.

In an all-electric storage drier the same drying load would require about 80 units of electricity, costing about 96p per ton. Costs will, of course, be higher if more than 6 per cent moisture has to be removed in adverse weather conditions; and for this reason it may be preferable in a large storage drier to use an oil-fired heat exchanger in conjunction with an electric fan. If this is done, running costs will usually be comparable with that of a continuous-flow oil-fired drier.

Storage driers are at their best in a warm, dry, early harvest, when practically no artificial heat is needed. In such seasons running costs are extremely low. Storage driers are least effective in districts with a cold, wet climate, when harvest may be very late, and very wet grain has to be dried, using atmospheric air that is cold and also has a high relative humidity. In such conditions a more rapid method of drying, using a continuous-flow or batch system is preferable; but even in such adverse operating conditions, running costs of a bulk drier are likely to be less than fixed charges for depreciation and interest.

Labour Requirement

On a large farm it is usually easy to justify the necessity of having a man continually on duty while a continuous drier is in operation. With smaller driers, the need for frequent attention is a disadvantage, so very small continuous-flow driers are of doubtful value compared with some forms of batch driers. A well-installed bulk drier has the advantage that it can be left unattended for long periods; but a ventilated bin plant suffers from the disadvantage of needing rather more labour for transferring grain from flat-bottomed silos if the grain has to be dried in shallow depths in a wet season, while an on-floor drier may need more labour when grain is loaded out of store. Self-emptying batch driers that are well installed, with separate wet and dry grain pits, can be very economical in the use of labour.

Labour required for operating ventilated bins has been reduced by the introduction of self-emptying 'air-sweep' floors, and labour requirements for attending to continuous-flow driers can be drastically reduced by the use of automatic controls. An automatic controller normally incorporates a sensing device which continuously measures the moisture content of the grain as it leaves the drier. If the moisture content changes more than about 0·3 per cent from the pre-selected figure, an electronic relay mechanism operates an adjustment which increases or reduces the rate of flow of grain through the drier. The provision of this equipment and a few other 'fail-safe' devices makes it practicable to leave a drier to operate unattended for long periods, and this, at a period of peak labour demand, can justify the cost of about £500 where the throughput is adequate, e.g. 500 tons per year.

Without automatic controls, the cost of labour may be reckoned at about 20p per ton for a large bulk drier and about 40p per ton for a small continuous-flow machine.

Total Drying and Storage Costs
The main problems in calculating farm drying and storage costs are the variable amount of storage provided by different installations, differences in the facilities for grain handling, and the amount of drying to be done in different seasons. Because of the differences in capital costs, a simple floor drier will almost inevitably show lower drying and storage costs than an installation employing a continuous-flow drier and bins, or even a ventilated bin plant. Table 33, which is based on typical installations chosen from Table 31, indicates that there is usually little difference in drying cost between the three plants, and that the total of drying and storage cost is closely linked to the capital cost of the drying and storage installation, in relation to storage capacity.

Table 33. Typical Costs of Grain Drying and Storage

Type of drier	Continuous, 6 ton/h, with bin storage	Floor-ventilated outdoor 100-ton bins	On-floor drying and storage
Tons per year dried and stored	700	500	800
Capital cost per ton stored £	45	20	30
Annual costs per ton dried and stored £			
Fixed costs *	6·30	2·80	4·20
Repairs and maintenance †	1·35	0·60	0·90
Fuel and power	0·65	0·96	0·96
Labour	0·40	0·20	0·30
Total cost per ton £	8·65	4·56	6·36

* Depreciation based on Table A.2(i) – assuming 200 hours' annual use of drying plant and life of 10 years.
  † Based on Table A.3, at 3 per cent of capital cost per annum.

Additional Costs of Drying and Storage

In addition to the costs already discussed, there are in some cases extra costs which would not be incurred if grain were sold immediately after harvest. These include extra drying, loss of weight, and loss of interest on value of the grain, or alternatively, loss of opportunity to invest in some other enterprise the money that will be realized by selling.

Extra drying

In some areas grain can be sold without penalty at a moisture content of up to 18 per cent. It cannot be safely stored in ordinary bulk stores at this moisture content, and so needs to be either further dried or kept in a store specially designed for 'chilling'. (See page 209.) The extra cost of drying from 18 to 14 per cent m.c. can be appreciable, e.g. 50p per ton for fuel and labour.

Loss of weight

The difference between selling at 18 per cent m.c. and 14 per cent m.c. is equivalent to approximately 0·05 ton of grain in every ton sold (see Table 34). The value will usually range from about £1·50 to £2·50 on every ton, and is therefore a quite important amount. Where grain is dried in bulk and stored at about 16 per cent m.c. the loss of weight by drying is halved, and this

Table 34. Approximate Loss of Weight in Grain Drying, Due to Evaporation Only

To calculate loss, multiply original weight by factor at intersection of initial and final moisture content columns.

Initial m.c.	Final Moisture Content. Per cent						
%	18	17	16	15	14	13	12
30	0·146	0·157	0·167	0·176	0·186	0·195	0·205
28	0·122	0·133	0·143	0·153	0·163	0·172	0·182
26	0·098	0·109	0·119	0·129	0·140	0·149	0·159
24	0·073	0·084	0·095	0·106	0·116	0·126	0·136
22	0·049	0·060	0·171	0·082	0·093	0·103	0·114
21	0·037	0·048	0·059	0·071	0·081	0·092	0·102
20	0·024	0·036	0·048	0·059	0·070	0·080	0·091
19	0·012	0·024	0·036	0·047	0·058	0·069	0·079
18		0·012	0·024	0·035	0·047	0·057	0·068
17			0·012	0·024	0·035	0·046	0·057
16				0·012	0·023	0·034	0·046

Example: 100 tons dried from 18 per cent to 14 per cent m.c. LOSS OF WEIGHT = 100 × ·047 = 4·7 *tons.*

is one of the potential advantages of a well-managed installation for drying and storage in bulk. Where the aim is to keep grain at this moderately high moisture content, care must be taken to see that it is cooled by ventilating with air that is both cold and dry.

Loss of interest on grain value or *loss of opportunity.* Receipts from grain sold can be used to reduce an overdraft, used as working capital for productive enterprises or invested outside farming. Interest at 8 per cent p.a. for a half year on a ton of grain worth £40 amounts to £1·60.

The sum of all the above costs, where all apply, may be as follows:

	per ton £
Depreciation and interest on a £30 per ton installation written off in 10 years	4·20
Drying and handling, say	0·50
Repairs and maintenance (complete installation) say	0·25
Loss of weight, based on 2·5% @ £50	1·25
Loss of interest for half-year at 8% p.a.	1·60
Total	7·80

It will be seen that a considerable extra return is required to pay for drying and storage costs. On the other hand, it should be remembered that depreciation and interest on capital represent a major part of the total, and that these costs can be lower with inexpensive installations, or where the installation costs have already been written off.

Returns from Grain Storage
Achievement of the potential returns from grain storage depend on selling the grain well, and taking advantage of the flexibility given by ability to store. A straightforward comparison between the total storage costs as calculated in the previous paragraph, and seasonal changes in the price of the crop, tends to over-estimate the expenses and to under-estimate the advantages of storage, because a farmer who cannot store is in a weak selling position, and often has to accept a price which reflects this. On the other hand, there is no positive guarantee that he will be able to realize to the full the benefits of seasonal changes in prices.

When storage capacity is available, it normally pays to make full use of it during the early months of the season; but the benefits tend to decrease as storage capacity on farms increases.

Grain Storage at High Moisture Content

For some purposes, and within certain limits, grain may be preserved and stored without drying. The main methods are (*a*) storage by exclusion of oxygen, in sealed silos, in unsealed tower silos or in plastic, elastic or similar flexible air tight materials; and (*b*) preservation by chilling to a suitably low temperature.

Storage in Sealed Silos
Barley for stock-feeding may be stored in many types of sealed silos. There is a wide range of equipment to choose from, ranging from galvanized steel silos emptied by a simple sealed auger, to vitreous-enamel-coated steel silos equipped with a 'breather-bag' and fitted for unloading by an effective sweep-

arm auger. The ideal grain moisture content for this form of storage is about 18–24 per cent. The method is suitable for stock feed only. Typical capital costs of installations with unloaders, before deduction of capital grants are:

Silo capacity tons	Cost of installation £
50	1500–1800
200	3200–3600
400	5800–7200

Such silos are likely to have a long life, but if written off in 10 years, annual fixed cost for depreciation and interest ranges from about £2 to £5 per ton. Operating costs are very low. This is a cheap method of storage compared with provision of a complete drying and storage installation; and results indicate that where equipment and management are satisfactory, losses are negligible and feeding value at least equal to that of a crop preserved by drying. There are, in fact, considerable advantages in high-moisture storage where barley is fed in large quantities to cattle.

Unsealed towers can give similar results, but are suitable only for large installations where there is a high rate of usage of grain, so that at least 3 in of grain is removed in a uniform layer daily. Capital cost of a large unsealed tower, complete with the necessary top unloader, is lower than that for a large sealed silo, ranging from about £13–£15 per ton before grant for a 400-ton silo. In smaller sizes, e.g. 200 tons capacity, capital cost is about £22 per ton before grant, i.e. higher than that of a sealed tower of similar size. An unsealed tower should only be chosen where it is certain that use of the grain will begin early in autumn, and will continue without any long interruption until emptying is completed. Moreover, the last of the grain should be used before the hot weather of summer.

Storage by Chilling
Grain having a noisture content up to about 20–22 per cent can be effectively preserved by chilling, using a refrigerator to cool the ventilating air below normal ambient temperature.

This method of high-moisture storage differs fundamentally from storage in the absence of oxygen, in that it can be applied, within limits, to grain required for any normal purpose, and not just for feeding. The temperatures recommended for preservation, in relation to grain moisture content, are given in Table 35. Chilling must be rapid if grain moisture content is over

Table 35. Recommended Temperatures for Chilled Storage
of Grain at Various Levels of Moisture Content

Grain moisture content per cent	Storage temperature			
	For no deterioration in 2 months storage		For freedom from mould, with slight fall in germinative energy	
	°F	°C	°F	°C
16	55	13	60	16
18	45	7	50	10
20	40	4	45	7
22	35	2	40	4

about 20 per cent m.c. It is therefore necessary to employ a unit of reasonable capacity in relation to the rate at which grain is brought in. One of the most economic types of installation is a large floor store, used in conjunction with a continuous-flow or similar drier, which is employed to reduce moisture content to the desired uniform level, e.g. 18 per cent m.c., before it is put into store. With such an arrangement, capacity of the chilling unit can be modest, and the complete chilling installation, consisting of the unit and simple portable ducting for distributing the cold air, need not cost much more than £1 per ton for capacities of 500 tons or more. Running costs of the chilling unit for such an installation can be very low, e.g. under 10p per ton for storage from September to March. Such plants are only likely to be attractive where there is an assured market for grain at about 18 per cent m.c.

Bulk Out-loading

The practice of transporting grain in bulk from the farm to the merchant, maltster or miller is making headway, and can often be economic, even on medium-sized farms. A number of firms supply prefabricated overhead self-emptying hoppers which will hold about 10 tons of grain and can be filled at leisure by the normal farm conveying systems. An alternative method favoured by many farmers is to provide a self-emptying silo near to the grain receiving pit and to install elevators of high capacity (e.g. 15–20 tons per hour) in such a way that a large load of grain can be quickly transferred from the holding silo to the road transport vehicle. Individual circumstances determine which method of out-loading is preferable, and it is a question that needs careful planning when a new installation is being designed.

Overall and Detail Planning Needed

A common planning fault is placing a drier or a grain store in an unsuitable building. With prefabricated silos which will carry a roof available, there is little to be gained by squeezing them into old barns: it is usually better to put them on a new site, which can sometimes conveniently be adjacent to the existing barn. It is often important to consider the requirements for efficient food grinding and mixing, and this will usually require the grain to be delivered by the normal grain conveying system to the hoppers of the crushing and/or grinding mill. Grain stores and food preparing plants are often excessively dusty due to bad planning and choice of equipment, and this is a fault that can be avoided at practically no cost if the problem is properly tackled at the design stage.

A common fault of construction with bulk grain stores is making deep grain receiving pits which are not water-tight. The advent of high-speed auger conveyors has made the construction of very deep pits largely unnecessary. Another way of getting round the problem without expensive asphalt tanking is the use of a pit of welded steel sheet of a standard size, which can be bought complete with elevator pit and dropped into the hole.

Harvesting Seed Crops

A full discussion of the harvesting of seed crops involves many technical details which are of greater importance than savings of cost. There is, nevertheless, a distinct trend towards low-cost seed harvesting methods wherever they can be applied, and many crops which only a short time ago were laboriously harvested by hand are now either direct combined or harvested by combine using the windrowing technique. Many farmers with a small staff can no longer find labour for the older methods, even if this results in a greater yield of good seed; and it is often a question of either using the combine harvester or abandoning growing the particular crop. Seed crops which are frequently directly combined include cocksfoot, fescues, red clover, sainfoin and linseed. Most seed crops other than cereals are windrowed before combining. Crops which can be successfully handled from windrows include ryegrass, white clovers, sainfoin, trefoil, timothy, mustard, rape, swede, turnip and thin-stemmed crops of sugar beet and mangolds. There are some crops which pay for stooking or tripoding where labour is available, and these include cocksfoot, fescues, timothy, sugar beet and mangold seed.

Most seed crops which are harvested by combine must be dried immediately after harvest at a temperature not exceeding about 110 °F (44 °C). Simple batch driers are usually suitable, but there are also one or two types of continuous driers which can be successfully used. The vertical tower type is generally unsuitable, and the most convenient are inclined cascade machines or oscillated semi-horizontal machines. Some conveyor driers will do the job but are difficult to clean out. Many seeds are difficult to convey because the seed will not 'flow'.

Harvesting Peas and Beans

Peas and beans are now normally harvested by combine. Beans are fairly easily combined direct provided that they are left until really ripe. They must then be harvested at times when the atmosphere is damp, in order to avoid shedding. Some drying is often needed, and this should be done slowly at low temperatures if splitting is to be avoided.

With peas, though the safest method for a high quality crop for human consumption is undoubtedly to use tripods, windrowing usually gives satisfactory results if done with care, provided the weather is reasonable. The crop is cut either by mower or, better, by a special pea cutter.

The peas are usually threshed much sooner from the windrow than from tripods, and a suitable drier is essential. An ideal type of drier for peas is one in which large quantities of only slightly warmed are are used – e.g. shallow ventilated bins, radial silos, tray and similar batch driers or floor driers. It usually pays to reduce the weather risks by combining as soon as the job can be done without damage to the peas. The cost of drying is increased, but the higher quality of the crop more than pays for the extra drying cost.

Harvesting Green Peas for Canning or Freezing

Green peas for canning or freezing are an important crop in many areas. The whole crop was formerly cut and transported for shelling by a stationary viner, preferably one situated within a few miles of the farm. This method is still used and is aided by advances in cutter-loaders and in transport arrangements, so that hand forking is no longer required; but haulage of the large amounts of haulm to central points is inevitably inefficient. So when effective mobile viners were developed their use quickly spread, in spite of a capital cost in the region of £20 000, exclusive of such necessary adjuncts as a well designed pea cutter-windrower, and also equipment for chilling the peas where the factory is not very near. Rapid harvesting when the crop is 'mature' is essential for maintenance of the very high quality demanded for processing.

All mobile viners in commercial use work on the windrow left by a pea-cutter. They are usually tractor-drawn, but self-propelled machines are now available. All have built-in automatic levelling devices, in addition to many easily made adjustments such as beater drum speed and beater angle. The shelled peas are handled to transport vehicles either in pallet boxes by means of a fork-lift, or by a hydraulically tipped dumping hopper. If the distance to the factory is considerable, the peas must be cleaned and cooled before transport. The common method of cooling is to immerse the peas in water which has been chilled, and pass the mixture through a cooling tower and over a grid to separate the water. Dry cooling has been adopted in some instances, and is likely to be further developed. The threshed haulm left behind a mobile viner may be picked up later if it is to be ensiled, or may be spread on the field if it is to be ploughed in.

Harvesting Dwarf Beans

Mechanical harvesting of dwarf, snap or bush beans has developed rapidly. Tractor-mounted machines have the picking mechanism at the front, and a pneumatic cleaning and collection arrangement at the rear. Other machines are tractor-drawn and self-contained. The trend is towards self-propelled harvesters. The beans are picked in a once-over operation, and transport is usually in pallet boxes. The bean harvester has revolutionized production of this crop, making it suitable for growing on a farm scale. As with peas, developments in mechanization have led towards closely spaced rows with higher plant populations and the replacement of after-cultivations by chemical weed control.

Harvesting Cabbage and other Above-ground Vegetable Crops

Cabbages and brussels sprouts on the stalk can be fairly easily harvested by machine. A cabbage harvester is normally a single-row side-mounted machine, which cuts the stalk, elevates by means of an endless conveyor specially designed for the job, and delivers to boxes or to a trailer. Efficient use of a harvester necessitates a once-over operation, and this means that the crop must be very uniform — usually a hybrid variety. It seems inevitable that more and more of the cabbage crop will be harvested in this way. Brussels sprouts for processing can be harvested in a similar way, the tops usually being left in the field, and the sprouts stripped off the stalks in a stationary machine.

For vegetable crops which necessitate selective harvesting, such as cauliflowers, there is no practicable alternative to hand cutting at present. There are, however, many ways in which hand work can be assisted. One method is to use large-capacity front-and-rear transport boxes on a row-crop tractor equipped with narrow-tyred wheels, which is driven through the crop with the cutting gang. Where possible, the tractor then runs its load to an indoor packing station.

Another method used for selective harvesting of cabbage, etc., in U.S.A. and Russia is to mount a long reversible conveyor-elevator on the rear of a tractor, and deliver the heads placed on it to a trailer. The endless conveyor consists of a horizontal central section and wings, either of which can be raised, according to need.

Mechanical harvesting of spinach for processing is a simple operation. The harvester consists essentially of a wide reciprocating cutter-bar, followed by an inclined full-width endless conveyor which moves rearwards, delivering the leaves to a cross conveyer which loads them into a trailer or lorry running alongside.

Use of machines such as bean, spinach and cabbage harvesters completely transforms vegetable production. It moves the crop from a category with a high labour requirement to one with a requirement for high capital investment in mechanized equipment.

CHAPTER NINE

Harvesting Root Crops

Potato Harvesting. Economic Aspects of Use of Harvesters. Systems of using Harvesters. Indoor Potato Storage. Sugar Beet Harvesting. Carrot Harvesting.

Potato Harvesting

Casual labour is still employed on many farms to assist in harvesting main-crop potatoes. Some forms of casual labour are satisfactory, but in general the use of such labour for this purpose is probably undesirable from a national standpoint, and is certainly a nuisance to many individual farmers. Moreover, obtaining casual labour becomes steadily more difficult and the cost continually increases. The problem of mechanizing the potato harvest on all farms is not, however, a simple one; for while there are now available complete harvesters which are quite efficient on suitable soils, these rather expensive machines fail to work efficiently in certain kinds of adverse conditions.

Area of potatoes		Percentage of holdings with potatoes	Percentage of total area of potatoes
acres	ha		
$\frac{1}{4}$–$9\frac{3}{4}$	0·1–3·9	73·4	20·1
10–$49\frac{3}{4}$	4–19·9	23·3	49·8
50–$99\frac{3}{4}$	20–40	2·5	17·4
100+	Over 40	0·8	12·7
		100·0	100·0

Thus, a machine which will work well on black fen with only two men on it may be unable to do a satisfactory job on wet heavy land with six operators on it.

The figures on p. 216 show size distribution of the areas devoted to maincrop potatoes on individual holdings in England and Wales in 1972, when the total area was 399 000 acres (161 000 ha), and the number of growers was 41 000.

When considering mechanization problems it must be borne in mind that the average area per farm is small, though a few large-scale farmers grow almost half of the total crop. The figures for sugar beet are similar.

Success in mechanical harvesting of potatoes depends to a great extent on adoption of a sensible policy regarding such matters as variety grown and time of maturity. As a general rule, potato harvesting in September is an easy job compared with harvesting in November. Moreover, harvesting a crop which is fully mature, and therefore resistant to mechanical damage, is easy compared with harvesting one which is not really ripe. Again, when soil conditions are reasonable, it is much easier to harvest a crop consisting of a relatively few large round tubers, which will roll on an inclined sorting table, than one with a lot of small tubers, of long or kidney shape. Little attention has so far been paid in Britain to choosing (and breeding) varieties to suit the limitations of mechanical separation methods.

Steady progress is being made towards the development of a complete potato harvester which can be relied on to work effectively and at a moderate speed, in any reasonable soil and crop conditions. Many modern harvesters can achieve a satisfactory performance in favourable conditions such as on light loams or peats, with only two operators on the machine to pick out a little rubbish from the final conveyor. Separation of soil and haulm are achieved by simple mechanisms, which work well provided that the soil is not excessively wet. There are, however, many soils, on which potatoes are regularly grown, which are either very stony or have a marked tendency towards cloddiness, or become very sticky when moist. On such soils the output of harvesters is low, and labour cost is relatively high, because forward speed is restricted by the need for workers on the machine to pick out every potato from the mixture of potatoes, stones and clods which passes through the harvester. In such conditions, a harvester with a gang of four working on it can often only achieve about $1\frac{1}{2}$ acres (0·6 ha) per day – a rate of work which may not be fast enough to get the job completed before conditions become so bad that the harvester cannot work at all. In such circumstances, many farmers who are able to get a gang to do hand picking prefer to do without the limited help that the machine can give. It has, however, been clear for a long time that hand picking will sooner or later become no longer practicable because labour will no longer be available for the work. Already, the cost of labour is a major factor which makes it extremely difficult for those who are unable to use a mechanical harvester to compete successfully with those who can. Gang-work picking costs can reach over £50 per acre (£125/ha) in a difficult season. Much effort is therefore being directed to the rapid development of harvesters which will sort mixtures of potatoes and stones or clods effectively. This is not, however, the only solution of the problem of the more difficult soils. Another is to take steps to improve soil conditions by pulverizing clods, removing stones, etc., so that the harvester has the relatively simple task of working in clod-free and stone-free conditions. Though it may at first seem fanciful to suppose that heavy soils can be made clod-free and stony soils made stone-free, it is in fact possible to bring about considerable improvements.

In heavy soils, the main method of improving harvesting conditions is to ensure that a fine tilth is prepared at planting time, and thereafter to take the utmost care to avoid any operation which will result in clod formation. This generally necessitates avoiding any deep intercultivation with tined implements. One method of approach is to use a herbicide spray, after covering the seed with at least 6 in (15 cm) of fine soil. Thereafter all cultivation is avoided if possible. Spraying tends to be somewhat more expensive than cultivations, a typical cost figure being a little over £4 (£10/ha) for spraying, compared with a little less for the cultivations which are replaced. The best technique is usually to wait until just before emergence, and then to use a finger weeder, light coverer and sprayer. Another method is to avoid cultivations likely to produce clods and to ridge up with a ridger which lifts a light sprinkling of soil gently on to the ridge. The 'Dutch' system, designed for heavy land, consists essentially of gradually building up a clod-free ridge by shallow cultivation and use of light coverer bodies.

A few farmers have tackled the problem of stones by separating them from the soil in the positions where the potatoes are to be planted, and putting them down again where the tractor wheels run. Another solution applied in some parts of U.S.A. is to use a strengthened harvester to deliver the stones into trailers, and to remove them for road-making. Methods which involve removal of the stones are inevitably expensive, and a great deal of work has been devoted to the development of harvesters which can separate stones or clods from potatoes during the harvesting operation. One method of separation in commercial use makes use of low-energy X-rays to distinguish between potatoes and stones or clods. When a potato passes through the beam between the source and detector, a finger deflects the potato on to a conveyor. When a stone or clod passes, the detector operates an electronic relay which causes the finger to move out of the way and allow the stone or clod to fall on to the ground.

A factor which helps to justify the high cost of such mechanisms is independence of casual labour. Whereas equipment which involves hand sorting can often only be used for about 6 hours a day, and for a 5-day week, harvesters with automatic sorting can often be operated for long hours, and if necessary at week-ends.

Economic Aspects of Use of Harvesters

Economic studies of potato harvesting make it clear that it is easy to justify use of complete harvesters where they can be made to work effectively. In comparing the results of using a harvester with hand picking, factors which must be taken into account, in addition to machine and labour costs, include the amount of leavings and the amount of damage to the crop.

Leavings
Wherever a harvester can be used, it scores over hand picking in respect of leavings. The amount of leavings with hand picking is very variable according to conditions, the equipment

used and the gang. In good conditions where an elevator digger is used, there is no reason why leavings should be high; but in poor conditions, using a spinner, it is difficult to avoid leaving almost 1 ton per acre (2·5 ton/ha). The average figure of $\frac{3}{4}$ ton per acre, found in a study in the East Midlands, is typical. With a well-operated harvester working in good conditions there is no reason why leavings should exceed $\frac{1}{4}-\frac{1}{2}$ ton per acre. Again, the average figure of the East Midlands study, of 0·35 ton per acre, is typical. This means that use of a harvester can be expected to result in an extra yield of about 0·4 ton per acre (1 ton/ha).

Damage

It is possible by mis-use of a harvester to do severe damage to the crop, which will result in a high proportion being unsuitable for the top grade. This, however, is avoidable, and is not characteristic of well-operated machines. Moreover, many hand-picking methods also result in a great deal of unnecessary damage. It is therefore impossible to give a comparative assessment which is of real value.

Labour

Labour requirements for harvesting are very variable, both for hand picking and use of a complete harvester. Labour for hand picking with a highly skilled gang using an elevator digger can be about 40 man hours per acre (100 man hr/ha) for the whole job of digging, picking, carting and storing; but such figures are typical only of areas where skilled workers are employed for picking. Where the gang consists of unskilled casual workers a typical figure for the same job would be double this – i.e. 80 man hours per acre (200 man hr/ha). This was the average figure found in an East Midlands study, and in earlier surveys in the North West.

The comparable labour requirement for using a complete harvester also varies with soil and crop conditions. In black fen conditions it can be less than 20 man hours per acre for the whole job of harvesting, carting and delivering to store; but a more typical figure is about double this, i.e. 40 man hours per acre (100 man hr/ha). (Average survey figures for manned machines include 40 in the North West and 41·5 in the East Midlands.) Thus, the average saving is about 40 man hours per acre – a figure which should increase rather than fall, as harvesters are improved. At a conservative estimate, this saving can be valued at £32 per acre (£80/ha); and this is a real saving, since there is practically never enough regular labour for potato picking.

Extra Costs of Operating Harvester

Extra costs of operating a harvester include fixed costs and repair costs of the harvester, and possibly a little extra tractor work owing to the slow speed of the harvester. Costs saved include repair costs on the digger, and it can be assumed that the latter will more than balance any extra tractor work. It can be assumed that a single-row harvester costs about £3000; and if it is allowed only a 5-year life, annual fixed costs are:

Depreciation	£600
Interest on capital	120
Total	£720

Repairs (Table A.3) can be reckoned at 5 per cent of purchase price per 100 hours of use.

A partial budget for harvesting 25 and 50 acres (10 and 20 ha) per year, on the basis of 2 acres (0·8 ha) per 8-hour day, is as follows:

	25 acres (10 ha) harvested £	50 acres (20 ha) harvested £
Extra returns and savings		
Extra yield at £12 per acre	300	600
Casual labour saved at £32/acre (£80/ha)	800	1600
	1100	2200
Extra costs incurred		
Annual fixed costs	720	720
Repairs and maintenance	150	300
Total	870	1020
Net gain from use of harvester	230	1180

On this basis the 'break-even' point at which it pays to use a harvester is below 20 acres (8 ha). Clearly, it would be economic to pay considerably more than £3000 for a harvester if this were necessary to give a higher level of efficiency, and ability to cope with a wider range of adverse conditions.

Systems of Using Harvesters

There are several distinct handling systems suitable for use with complete harvesters. The most common is side-delivery from the harvester into a tipping trailer which runs direct from harvester to the field clamp or store. To avoid holding up the harvester, the minimum transport gang is two men, two tractors and two trailers. Three may be necessary for a high-output harvester working in good conditions at a long transport distance.

Where the transport distance is more than a mile or two, e.g. where the crop is taken to a central grading and packing station, the trailers which work in the field may transfer their loads via an elevator to a large-capacity self-unloading body which is set down at the road-side and later picked up by an articulated truck. This double handling can cause extra mechanical damage unless special care is taken to avoid it, but in wet conditions it has the advantage of removing a considerable amount of soil.

Where the crop is to be stored in large (e.g. $\frac{3}{4}$ to 1 ton capacity) pallet boxes, these may be filled in the field by placing two or more on a trailer, which is drawn alongside the harvester. One of the potential disadvantages inherent in this method is the damage that may occur in delivering the tubers from the end of the harvester elevator into the bottoms of the boxes. It is difficult, even with a good hydraulically operated extension on the harvester elevator, to avoid dropping the tubers farther than is good for them. A good solution of such problems clearly depends on carrying the box either on the harvester or on the tractor which tows it. This solution is equally desirable from the viewpoint of saving labour. Where a trailer is drawn alongside a single-row harvester, a man with tractor and trailer typically spends almost an hour in collecting a load of 3 tons. There are therefore considerable advantages to be gained from effective systems of collecting the crop in bulk hoppers on the harvester. Mechanisms which automatically raise or lower the delivery end of the loading elevator to ensure only a short drop into the transport vehicle should easily justify their small extra cost.

The system of drawing a bulk vehicle alongside the harvester comes into its own where the multi-row harvester has a high output.

Mechanical harvesting is easy on those soils such as black fen, coarse silty loam and light sandy loam, where clods are no problem and there are few or no stones. In Britain, only a small proportion of the crop is grown in such ideal soil conditions. The difficulty of separating potatoes from stones and clods is a very real one, and there is the additional trouble that avoidance of bruising or rubbing of the skins is not easy where separation conditions are adverse. A modern potato harvester is a good deal more difficult to set than a machine like a combine harvester, which after being set in a given crop is usually able to make a fair job, even if the operator does not pay much attention to correct settings. As with elevator diggers, the general aim should be to maintain a cushion of soil with the potatoes until the point where severe agitation ceases.

Since some of the crop must continue to be harvested by spinners or elevator diggers, it is worth while to consider means of saving labour when such machines are used. One system which is widely used is the continuous 'stint' method which involves picking into baskets and transporting in bulk. Disadvantages are the need for a fair-sized gang, two tractors and trailers, and a large number of rather expensive picking baskets. This system can be efficient in terms of labour requirements, but it involves a great deal of very hard work on the part of one or two men who throw the full baskets into the carts or trailers. It also needs at least 8 pickers in order to save wasting an undue amount of the potato digger's time. Time studies have shown that the whole operation of emptying the field line of potato baskets into the trailer only needs about 10 to 12 man minutes per ton, but a frequent consequence of such a working rate is over-strain of a skilled worker.

With a small gang, a better method (sometimes called the 'breadth' system) is to lift several rows in advance, using a 2-row elevator digger or an elevator digger with offset delivery. A picker then takes a single or double row, and the trailer works down the rows, keeping just behind the gang. This method is economical in baskets and mechanical equipment, and time studies show that it is the best way to use a small gang.

Efficiency of a potato-picking gang depends to a large extent on provision of a good system of transport from field to store. Picking into bags results in waste of time in emptying baskets into the bags, followed by the slow and laborious operation of lifting bags from the ground and loading them on to trailers. By comparison, picking into pallet boxes has many advantages. Emptying baskets into the boxes is a straightforward job, and handling the boxes from field to store can be done quickly and without any hand work by using a tractor loader.

Where the only function of the pallet box is to assist in collection and transport of the crop, a convenient capacity is $\frac{1}{2}$ ton. There are three main ways of using the boxes. With a large picking gang working on the 'stint' system, each picker or pair usually has two boxes, though one each and a few spares may suffice with some bulk transport systems. Provision of enough boxes avoids any hold-ups due to failure of transport, permits transport in the boxes if desired and makes it easy to operate a piece-work method of payment. Where transport is in bulk trailers, a tractor loader fitted with a special box-handling head picks up the full boxes, tips them as gently as possible into the trailer and sets the empty boxes down where next required by the pickers. With a large gang and a long transport distance, the transport gang usually consists of one man with a loader and two with tipping trailers. The loader operator has some spare time, which he spends in moving partly-full boxes to reduce the carrying distance. As a general rule, the aim should be to avoid having to carry full baskets more than about 50 ft (15 m). A box-emptying device with good hydraulic controls to provide a gentle tip makes it possible to avoid excessive bruising. For smaller picking gangs bulk transport systems operated by only one or two men can easily be devised. The smallest consists of one man with tractor loader and a trailer which is unhitched and set down during loading. Minimum loading time for this one-man system is about 10 min per ton.

An alternative method of transport is to load the boxes on to the flat bed of a trailer, transport the load of full boxes to store, and there tip direct on to the heap, preferably using a loader with a hydraulically controlled extension to give extra loading height. This is an attractive system in that it avoids the cost of an elevator, and two extra handlings.

Where clamping is in the field, or the store is very near, the boxes may be transported on special carriers on the front and rear of the tractor. This is a good method for use in conjunction with a small gang working on the 'breadth' system of picking. The tractor with one box at the front and two at the rear moves along with the pickers and can keep the boxes to be filled very close to the work. Moreover, the low height at which baskets are emptied makes the work easier than when a trailer is used. The tractor driver can help with emptying of baskets.

For early and second-early crops it is often an advantage to bag off and weigh on a platform on the harvester. Handling is facilitated by provision of a spacious platform, a weigher with devices to assist rapid adjustment of weight and pallets for subsequent handling of the bags. After a pallet has been loaded on the harvester, it may be set down by a simple hydraulic device, ready for handling by a hydraulic loader and trailer or truck.

Indoor Potato Storage

Bulk potato storage in buildings has rapidly become widely accepted. Advantages compared with the old method of storing in small clamps covered by straw and earth may be listed as follows:

(1) No necessity for long straight straw, which is difficult to obtain when combine harvesters and pick-up balers are used.

(2) Clamping and earthing up eliminated.

(3) Opening, uncovering and re-covering of clamp eliminated.

(4) Good working conditions for grading and bagging, even in bad weather.
(5) Ability to riddle at any time allows work to be planned in advance and secures better overall labour organization.
(6) Increased riddling outputs of up to 50–100 per cent.
(7) Loading under cover, and no difficulties from inaccessibility to road transport vehicles.
(8) Less loss from disease due to drier store conditions.

Control of chitting by the use of low-temperature ventilation and/or application of chemical sprout suppressants makes it possible to keep potatoes in good condition in late spring and early summer, after this is impossible in the old type of clamp.

The disadvantages are:

(1) Transport to the store may require an extra tractor and trailer owing to the longer distance.
(2) Some small extra loss of weight (about 2 per cent) in store due to drier conditions.
(3) Cost of the store and of loading equipment.

Though it is clearly possible to develop methods of operating indoor stores so as to avoid extra loss of moisture, a criticism of potatoes from most existing stores is that they usually lack the 'bloom' of potatoes kept in clamps. At the end of the season this can result in a considerable price difference for 'pre-packed' coloured potatoes.

This latter criticism in particular, together with the cost and transport aspects, has led to a search for alternative methods of clamping which are more in line with modern needs. One result has been the adoption of a technique of using very large clamps, walled by straw bales and polythene, and fitted with ventilating ducts and chimneys. Such clamps, with a bottom width of about 15 ft (4·5 m), walls 6–8 ft (1·8–2·4 m) high and a pair of A ducts running parallel to the walls, have a capacity in the region of 10 tons per yard (0·9 m) run, compared with about 1 ton per yard run of the old type of clamp. The ventilating ducts must be well made and adequate in size.

Most existing farm buildings can be easily converted to use as bulk potato stores, and as this can usually be done at a cost of no more than about £2 per ton stored, this is the method to adopt where a suitable shed or barn is available. There are three points, however, which must be carefully studied, viz. (1) strength of the walls, (2) protection from frost, and (3) ventilation. Any farmer who is seriously considering the question is well advised to study an official leaflet covering all aspects, but a few of the main points are briefly outlined below.

Potatoes stored in bulk have an angle of repose which varies according to shape and cleanness of the tubers, but averages about 35°. When the crop is stored in bulk between vertical retaining walls there is a considerable side thrust on the walls, which increases with storage depth. Existing barns with strong brick walls may be quite safe at shallow storage depths, but farmers should seek advice on the safe limits and reinforce if necessary. Insulation from frost is most simply and cheaply provided by a single layer of baled straw all round the insides of the walls. Even in a new building this is often as good a method of insulation as any, but if the maximum use must be made of a limited area it is possible to insulate satisfactorily inside brickwork by (1) painting the walls with a coat of bituminous paint to keep out the damp, (2) fixing to the walls thick boarding made of cement-bonded wood shavings, and (3) cement rendering the inside. The boards must be nailed on using large washers, so as to make an air space between the board and the brickwork.

For other types of walls, as well as for brickwork, there are many proprietary insulating materials to choose from.

If mature crops are stored in a clean and dry condition in well ventilated buildings at depths not exceeding 8 ft (2·5 m) it is not essential to provide any bottom ventilation, though it is always safer to do so just in case of trouble developing due to immaturity or disease. In all cases where potatoes are stored from wall to wall at heights above about 6 ft (1·8 m) it is advisable to be on the safe side and provide for bottom ventilation by means of triangular slatted ducts which are laid on the floor at intervals of about 6 ft during filling, and run either from the outside towards the centre of the store, or from a central underground duct towards the outside. It must, of course, be possible to close all ducts completely during frosty weather.

The potatoes should always be covered on top by a layer of loose straw about 2 ft (0·6 m) deep. This avoids having moisture condense in the top layer of potatoes and at the same time provides protection from frost and prevents greening by keeping out light.

A simple store which is filled from wall to wall to a height not exceeding 8 ft (2·5 m) can be fairly easily loaded using a normal type of engine-driven general-purpose elevator, with a hopper specially adapted for potatoes. Where an existing building is to be used, therefore, a decision to try indoor storage involves practically no capital cost; and since there is advantage in storing a proportion of the crop indoors, even if the whole cannot be so handled, there is every incentive to make use of such buildings as are available. In the main potato growing areas, however, there are often no suitable buildings available for conversion, and a decision to use indoor storage involves putting up a special building. Here it will probably pay to store at a depth of about 12 ft (3·5 m) and adequate forced ventilation must be provided. The net capital cost of such a building, after deduction of grant, should not exceed about £12 per ton stored – and it can be less on farms where the bulk potato store is housed under the same roof as a bulk grain store.

There are many reasons for limiting loading height to about 12 ft (3·5 m), one of the chief being the difficulty of loading at greater heights. Even at this height, satisfactory loading calls for a specially designed elevator which is inevitably fairly costly if equipped with the necessary devices for removing loose soil, and with a swinging extension at the top to facilitate even loading and the avoidance of soil cones.

At any depth over 8 ft (2·5 m) forced ventilation is essential for safe storage in all circumstances. The objects of ventilation in British conditions are (1) to dry off quickly any parts of a crop that are put into store in a wet condition; (2) to dissipate the heat developed by respiration of the tubers. This is the main function of forced ventilation. (3) To avoid the spread of rotting.

Methods of managing a store immediately after loading need careful study. As a general rule, it is considered advisable to allow a sound crop to warm up, the aim being to maintain a temperature of about 60 °F (16 °C) at first. This is called the 'curing' period. The object is to encourage the formation of corky tissue on wounds caused during harvesting. After about 10−14 days the temperature should be reduced as quickly as possible to about 45 °F (8 °C), but with as little blowing as possible. This means that unless rapid heating is taking place, a little ventilation in cool weather conditions is all that is needed.

When potatoes go into store wet, immediate drying is recommended, and then the store should be cooled as quickly as possible to reduce danger of rotting beginning at the points of contact, which cannot be effectively dried.

Ventilation equipment for British conditions is usually designed to provide about 40 cu ft per min (0·02 m³/sec) per ton stored − for half of the total storage capacity. Sprouting may be prevented by introducing a volatile sprout-inhibiting chemical (e.g. Nonanol) into the ventilation system when growth is just starting. It is, of course, essential that the ventilation be easily controlled in separate blocks, but there is considerable advantage in a system which enables the full fan capacity to be concentrated on a trouble spot, as opposed to one where each block has an independent fan which cannot ventilate any other section.

Good top ventilation of a potato store is quite essential, since lack of it may result in the potatoes being wet on top − a condition conducive to rapid rotting. Where ventilation conditions are good, on the contrary, the top cover when removed provides good litter, while the baled straw walls can be used for several years.

In planning a new store it is advisable to allow a minimum width of about 24 ft (7 m), and to arrange to leave a space of about 20 ft (6 m) which will not be filled at the main unloading end. This facilitates inspection, and storing of equipment such as the loader and sorter, while it allows riddling to begin under cover without any difficulty. In a large store consisting of a centre section and a lean-to on each side it is sometimes advantageous to be able to erect removable dividing walls between the three main sections. If this is done, it is usually best to provide independent access to each main section. In a large store it is always advisable to provide for emergency unloading from the end that is normally unloaded last.

Where potatoes are stored for crisping, it is considered by the processors that the crop should be stored at not less than

45 °F (8 °C) in order to avoid the formation of reducing sugars, which cause darkening when the potatoes are cooked. However, some of the adverse effects of storing at lower temperatures can be removed by giving the crop a period of storage at about 60 °F (16 °C) before processing. There is still much to be learned about the best management methods, but it is at least certain that a store for crops intended for processing should provide for a fine degree of temperature control and also for the introduction of a chemical sprout suppressant.

Some stores have been designed for storing the crop in 1-ton boxes. This eliminates the need for reinforced walls, but necessitates a good system of insulation and ventilation, with the possibility of recirculation of air when necessary. Capital cost for a small store of 500 tons capacity is likely to be in the region of £10–£15 per ton. The boxes themselves can be farm-built at a cost of about £6 per ton.

There is little to be gained by attempting to make a precise estimate of the costs of and returns from particular methods of indoor storage compared with the clamp method. Most of the important considerations have already been mentioned, and it is only in such matters as reduced riddling costs that the savings are easily calculated. Potato growers will readily understand that savings in labour are unimportant compared with such factors as ability to sell when prices are high, and ability to ensure that in a season when keeping quality is poor a higher proportion of the crop can be saved from deterioration.

In estimating annual costs of owning and operating a new store, it can be assumed that a well-constructed building and electrical equipment will have a physical life of 15 years, but it may be desirable to assume a 10-year life for calculation of depreciation. Annual fixed costs may therefore be reckoned as follows:

	Annual Depreciation and Interest, £ per ton	
	10-year life	15-year life
Building at £8 per ton	1·12	0·85
Building at £12 per ton	1·68	1·28

The life of mechanical equipment can reasonably be estimated at 10 years. Capital cost may be of the order of £4 to £8 per ton stored, giving annual costs of £0·56 to £1·16 per ton.

Running costs are normally low, even in a temperature-controlled store. This cost should not normally exceed about £0·50 per ton.

Total annual cost of a purpose-built indoor store is therefore likely to range from about £2 to £5 per ton. This is not a high cost to pay for the undoubted benefits, but is a factor in causing farmers to look for alternative methods which involve lower capital costs.

The returns from potato storage fluctuate widely with the season. There are times when the greatest profit can be made by selling the crop directly from the field; but farmers who rely on potato growing year in and year out must be able to store maincrops as a general rule. The market price usually tends to rise between mid-October and March by about £5 per ton in a 'surplus' year, and by well over double this amount in a 'shortage' year.

Sugar-Beet Harvesting

The mechanical harvesting of sugar beet represents one of the major triumphs of mechanization in farming, and serves to indicate the progress that may be expected in future with many other difficult farm jobs. In the early 1930s mechanical harvesting as a regular method seemed a long way off. The standard of topping insisted on by the factories and the need to avoid wastage seemed to make complete harvesters a vision of the distant future. In fact, however, with many countries working on the problem, many widely differing machines were soon developed. Today the whole of the crop in Britain is mechanically harvested.

A typical modern complete harvester either delivers the topped and cleaned roots directly into a vehicle running alongside, or collects them in a self-unloading tank on the harvester. It therefore saves loading, as well as the operations of loosening, pulling, knocking and topping. The life of a modern harvester is satisfactory. One manufacturer of a machine equipped throughout with sealed bearings estimates its life to be at least 2500 hours – i.e. 1000 hours more than the best of the conservative figures for beet harvesters in general suggested in Table A.2(i).

Technical aspects of a sugar-beet harvester's performance can have a considerable effect on financial returns. A poor harvester may leave a considerable weight of roots on or in the land. Checks made at harvester trials usually show a range of total losses from about $\frac{1}{4}$ to 1 ton per acre. Dirt tare and under-topping together often range from about 8 to 18 per cent, and can result in serious deductions; while over-topping can be another serious source of loss. The best machines will give a poor technical performance if they are not properly adjusted.

Tanker harvesters have considerable advantages over 1-row side-loading machines, and may be expected to supersede them. A harvester without the ability to carry the roots which it harvests inevitably requires a gang of at least two men before it can begin work. This is a serious limitation. A tanker harvester, on the other hand, can work in a variety of ways to suit availability of labour. If necessary it can be operated on a one-man system, collecting a tank-full of beet, and running with the load to a heap in or near the field. Transport and unloading naturally occupy a considerable amount of each working day, and average output of a single-row harvester on this system is only about $1\frac{1}{2}$ acres (0·6 ha) per 8-hour working day. However, studies in the Eastern Counties have shown that this

Table 36. Comparison of Sugar-Beet Harvesting Systems in the Eastern Counties

| | System | | |
	1-man	2-man	3-man
Type of harvester	Tanker with elevator	Side-loading	Side-loading
Number of tractors	1	2	3
Number of trailers	0	1	2
Average area harvested			
per 8-hour day, acres	1·9	1·3	1·7
hectares	0·8	0·5	0·7
Average man hours			
per acre	4·3	12·7	13·9
per hectare	10·6	31·4	34·3

one-man system can approach theoretical output rates, in a way that larger gangs never can. Comparisons in this study gave the results shown in Table 36. The one-man system, as so often happens, was more efficient in every way. However, in order to get maximum output from an expensive harvester it may be desirable to operate a 2-man system. The trailer should preferably be a high-tipping type, or one with a power-operated mechanical unloading device. The harvester works more or less continuously, unloading its tank into the trailer while on the move. With such a two-man system an overall working rate of 0·3 acres (0·12 ha) per hour should be attained in good working conditions, and operating cost per acre can be estimated as in Table 37. In practice, it would be difficult in most areas to maintain this working rate right to the end of the season; so the tractor and labour figures for the larger areas may be too low. The 2-man tanker system compares badly in efficiency with the 1-man system; so it should only be used where there is real advantage in a slightly higher overall harvesting rate.

Table 37. Cost of Harvesting Sugar Beet in Good Operating Conditions (2 Men) Using Tanker Type Harvester Costing £1200

		Area Harvested Annually			
	acres	15	30	60	90
	ha	6	12	24	36
Estimated hours of machine use [1]		50	100	200	300
Annual fixed costs £					
depreciation [2]		120	133	200	240
interest		48	48	48	48
Total annual fixed cost £		168	181	248	288
Annual repair costs [3]		36	60	84	108
Total annual machine cost £		204	241	332	396
		£	£	£	£
Machine cost per acre		13·6	8·0	5·5	4·4
Tractor cost per acre (2 tractors)		4·0	4·0	4·0	4·0
Labour cost per acre (2 men)		5·3	5·3	5·3	5·3
Total cost per acre		22·9	17·3	14·8	13·7
Total cost per ha		57·3	43·3	37·0	34·3

[1] See Table A.1(iv).   [2] See Table A.2(i).   [3] See Table A.3.

With side-loading harvesters there should be every advantage in the use of multi-row machines. Two-row and 3-row machines are the accepted type in many European countries, and it may be expected that multi-row side-loading machines will continue to compete with single-row tankers in Britain. Any side-loading harvester requires a minimum gang of 3 men and 3 tractors for continuous working; and even with such a team, the average rate of work of a single row machine tends to be appreciably less than that of a single-row tanker and a gang of two. Five-row or six-row, 3-stage harvesters have a high potential work rate and seasonal capacity; but a large gang and very good organization of transport are needed to secure a high output. Work naturally tends to fall to the rate of the slowest unit, which tends to be the loader, because it has to wait for trailers. However, topping and lifting units may be able to leave off earlier in the afternoon. The work of 3-row complete and 2-stage systems tends to be easier to organize.

High forward speeds with most types of machines tend to lead to high losses. In a typical trial, losses which were already too high at 1·6 ton/acre (4 ton/ha) at 3 m.p.h. (4·8 km/h) rose to 3 ton/acre ($7\frac{1}{2}$ ton/ha) at 5 m.p.h. (8 km/h).

It is no longer worth while to consider hand harvesting of sugar beet. The operations of pulling, topping and heaping alone require about 50 man hours per acre, compared with the 7 man hours per acre for the whole job of harvesting and transport from the field by two men using a tanker harvester. Even at a ridiculously low annual usage of 15 acres (6 ha) per year, an expensive tanker harvester is a better proposition than hand work at the normal piece-work rates. In an Eastern Counties investigation, where the harvesters were mainly simple side-loading types, it was economic to own a harvester for use on only 4 acres (1·6 ha) if opportunity costs were not considered. When the acreage to be harvested is low, serious consideration should be given to opportunity costs. Opportunities missed by owning a harvester may include such factors as (a) other use of the capital for productive purposes such as grain or potato storage, and (b) other use of labour for such purposes as drilling winter wheat in good time.

Beet harvesting is an operation which is very suitable for contract work. A good contractor should be able to get the work done according to a pre-arranged schedule; and shortage of capital may make this a good solution for the farmer. Machinery syndicates are equally attractive. Considering a typical case of neighbours with 40–50 acres (16–20 ha) between them, a good tanker harvester shared on a syndicate basis is likely to be a far better proposition for both than if each looks after his own needs by buying a cheap side-loading machine.

Collection of tops for feeding is catered for in a variety of ways. Some complete harvesters collect the tops in a container and windrow them at right angles to the work. Other put 3–6 rows of tops into a single row parallel to the rows of beet. Other methods of immediate collection include direct loading into a trailer running alongside the harvester – but this makes the whole job of transport at harvest an extremely big undertaking which cannot be tackled on most farms. One other method of top harvesting that has been tried is the use of a direct-cut forage harvester to collect the leaves before using a harvester fitted with a topping device. The disadvantage of this is that the leaves without the crowns do not constitute such a generally useful food as when the tops and crowns are harvested together.

Dirt tares may be high if beet harvested in wet conditions are delivered direct to the factory. This difficulty can be effectively overcome by clamping the crop for about a fortnight and using an elevator fitted with an effective cleaning mechanism to load the crop on to the road transport vehicle. Many farmers have come to the conclusion that it pays to lay down a concrete apron at well selected points to facilitate re-loading by means of a front loader (Chapter 4).

A disadvantage of mechanical harvesting on heavy land is the necessity to complete the field work fairly early in the autumn. This results in some reduction in yield compared with crops that are allowed to continue to grow for an extra month.

Farmers also need to guard against losses of beet in the harvesting operation, since experimental work shows that these can be appreciable with badly operated machines in wet conditions.

One of the chief remaining difficulties in beet harvesting is the amount of damage done to soil structure by the tractors and trailers used in harvesting and carting off the crop. With a single-row harvester every single inter-row space right across the field is first pressed down by the tractor and harvester, and then by the tractor and trailer running alongside. If harvesting takes place in wet conditions the result can be detrimental to subsequent crops. Where 2-row or 3-row harvesters are used, though the machines themselves are heavy, considerably less soil is compacted or smeared by tractors, harvesters and trailers.

Preparation for mechanical harvesting, as for precision drilling and mechanical thinning, begins in autumn with level ploughing. Drilling must be straight, and the field must be laid out to facilitate use of the harvester and avoid unnecessary hand work at row ends.

Mechanical Carrot Harvesting

A common method of harvesting carrots is to loosen the row by use of a short-tined hand fork; pull out a handful by the tops; wring off the tops using both hands; and sort the roots into market and processing grades. This method is preferred by some growers for a high-quality trade; but the public preference for washed carrots and the high labour requirement of the hand method make it fairly certain that this method will soon disappear.

A study in the East Midlands showed that hand work needed 8·8 man hours per ton at hourly work rates and 6·3 man hours per ton for piece work.

Many carrot harvesters top the roots while they are still in the ground; lift them by use of inclined shares similar to those used in sugar-beet harvesters; raise them and remove loose soil by means of a closely spaced elevator chain; and after providing for some hand sorting by workers riding on the machine, deliver the carrots either to a bagging platform or to a trailer drawn alongside. The topping in this type of machine takes place well ahead of lifting, and may be carried out by steel blades revolving about a vertical axis, rubber flails revolving about a horizontal axis, or a combination of these. This method is suitable for single rows, or for 2–3 rows drilled fairly close together, with fairly wide spaces between the multiple rows.

Another type of harvester is designed for lifting the whole of a bed of 12 rows, 4 in (10 cm) apart, using a share and elevator chain similar to those employed on a 2-row potato harvester. In this case topping is a separate operation, and is carried out by a suitably modified flail forage harvester which exactly spans a single bed, and blows the pulverized tops on to the land which has already been cleared. Very level beds and careful setting are of course essential for satisfactory topping. The main reason for choosing the bed system, which tends to make harvesting difficult, is a cultural one. This system gives the maximum yield of relatively small carrots of a uniform size.

A third type of harvester works like some sugar-beet and beetroot harvesters, which lift the crop into the machine for topping. The tops are gripped by a pair of rubber belts, which pull the roots out of the soil as they are loosened by inclined shares. In the harvester there are devices for topping at a fixed height from the crowns. Cuts, bruises and broken tips all detract from the value of the product, and such mechanical damage is minimized with this type of mechanism. The belt type of lifting mechanism shows to advantage on stony soils, since any type of chain-link lifter web tends to raise stones as well as crop: this can cause serious separation and transport problems, and also inevitably leads to bruising damage.

As regards economic aspects, a typical harvester requires only 5–6 machine hours per acre (12–15 hr/ha), and the saving of labour is so great that the cost of a fairly expensive harvester can be justified on as little as 4 acres (1·6 ha) of crop. Technical faults in harvesting can be partially remedied during washing and grading; but the proportion that is marketable may be considerably reduced. Output of the harvester tends to be limited by the washing and grading process, and by market requirements.

CHAPTER TEN

Livestock Engineering

Capital Investment in Equipment and Buildings. Building Layouts. Transport. Grinding and Mixing Feedingstuffs. Livestock Feeding. Layouts for Self-Feeding Silage and 'Easy-feeding'. Mechanized Cattle Feeding Systems. Economics of Mechanized Cattle Feeding. Mechanized Pig Feeding. Electric Fences. Cleaning and Littering. Milking. Portable Milking Plants. Labour for Milking. Choice of Milking Parlour. Milking Machine Management.

Mechanization of agriculture in Britain was at first concerned mainly with the application of mechanical power to field operations. By the 1950s this had reached a high level of development, and attention was increasingly turned to mechanization of work in and about the farm buildings, a high proportion of which is concerned with livestock tending. The importance of various livestock products in farming is indicated by the figures given in Table 3, which show that livestock and livestock products represent about 70 per cent of the total agricultural output.

Intensive methods of production have become essential for commercial success in production of milk, pig-meat, eggs and table poultry; and though production of beef and sheep from grass on hill farms remains to some extent independent of mechanization developments, advances in methods of grass production and conservation, and in transport and handling, inevitably affect an increasing proportion of these.

Standards of management of livestock enterprises have already made substantial advances – so much so that figures for labour requirements for stock husbandry which were reasonable a short time ago have already been made completely obsolete so far as many highly mechanized enterprises are concerned.

Modern figures for reasonable labour requirements for various types of livestock are given in Table A.4 and A.5. 'Livestock Engineering' is a term which may be applied to the many developments, including automation, which are designed to make drastic cuts in the time devoted to such jobs as feeding, cleaning and littering. Other important items are milking, which typically occupies about 40 per cent of the time devoted to dairy cows; and handling of eggs, which accounts for a high proportion of the time spent on laying birds when feeding and cleaning are highly mechanized.

Capital Investment in Equipment and Buildings

One of the major problems in livestock engineering is the capital required to introduce an efficient installation. When substantial improvements are required it is rare that the existing buildings are really suitable; and the problem of whether to make do with the old buildings or to erect new ones has to be considered. No attempt can be made here to study in detail the economics of providing new buildings, and reference to other sources of information [22] is recommended.

Though buildings may have a long physical life, it is normally necessary in making management calculations to write them off within a reasonably short period, in order to ensure the return of capital for re-investment if desired. As a general rule, the life assumed should not exceed 15 years, and for many purposes 10 years is a more business-like assumption. If a 10-year 'life' is assumed, annual fixed costs on every £1000 spent are approximately as follows:

Depreciation	£100
Interest, maintenance [22] and insurance	50
Total	£150 = 15 per cent per annum

This may be reduced by grant under the Capital Grants Scheme.

Building Layouts

Before studying details it is worth while to consider broadly the planning of building layouts and materials handling. It is seldom that a farmer has the opportunity to plan a new building layout in its entirety, but the opportunity does sometimes arise to make advantageous changes in the use of existing buildings, which in many cases were built at a time when the needs were quite different from those of today. Points of general application, some of which have already been detailed in Chapter 4, are as follows:

(1) Where possible, have the cowshed or collecting yard easily accessible to the main roadway from the pastures. This helps to avoid having unnecessary gates and a muddy road through the buildings.

(2) Stack hay and straw as near as possible to where it is to be used, and preferably right alongside. If it is not possible to stack hay in a position where it is easily accessible for hand feeding it is advisable to have temporary storage for a trailer load near to the feeding point in order to avoid long journeys with small loads to and from the main store.

(3) Plan silos and methods of feeding silage so as to eliminate carrying small loads into sheds or yards.

(4) Keep resting and feeding areas separate in order to make litter last longer and to keep stock comfortable.

(5) Design or alter buildings where practicable to make use of mechanical equipment for handling farmyard manure or slurry.

(6) Eliminate all unnecessary movement of materials as, for example, from grain store to food preparing plant when home grown grain is ground and mixed on the farm.

Some of these points are referred to again below.

Transport

A brief consideration of the scope of farm transport emphasizes the importance of carrying out the many transport jobs with a reasonable degree of efficiency in labour utilization. Among the jobs which consist almost entirely of transportation are the feeding of all livestock, the cleaning of cowsheds and the moving of milk from cowshed to dairy.

These jobs are as important on many farms as the operations more directly associated with crop production such as taking fertilizers and seeds to the field, and carting off the harvest of arable and fodder crops.

The difference between a good and a bad layout and scheme of transport is well illustrated by reference to the task of feeding a hundred pigs. With a well-laid-out piggery having a central feeding passage, water laid on, a smooth short concrete path leading from the food store to the troughs, and a large-capacity rubber-tyred feed barrow, the whole job can be done by one man in about 10 minutes. If, on the other hand, the food has to be mixed wet, and carried in buckets along muddy passages to inaccessible troughs through pens of hungry pigs, the job may take well over an hour. This is a job that has to be done morning and night every day, and the saving in labour cost of a good layout and good equipment over a year is enormous.

The laying of concrete is a simple job, especially if a concrete mixer is used. Tractor-mounted p.t.o.-driven concrete mixers are now available and may be a worthwhile investment where there is a large amount of work to be done over a long period. Where the job can be quickly done, on the other hand, hiring a mixer will be preferable. A still better proposition for most farmers is to buy ready-mixed concrete. This is now widely available, and its use not only saves the labour involved in mixing, but also ensures a degree of reliability and uniformity in the product.

On many farms, once smooth concrete roads and passages have been made where transport is regularly required, no special vehicles are needed beyond hand-pushed rubber-tyred trucks and the normal farm tractors and tipping trailers. A 4-wheeled hand truck with a flat platform is suitable for moving small loads of baled hay or straw, or for feeding by hand fairly small amounts of silage. For regular use in feeding meals or dairy cubes, a hopper type hand truck with a pair of main wheels on a central axle, and a small wheel at each end is handy, being very easy to turn.

Where the transport problem is sufficiently regular and standardized, and size of the enterprise justifies the capital cost, complete mechanization by means of automatic devices may be worth consideration. Cafeteria laying batteries with automatic feeding and cleaning are now commonplace. Automatic feeders and mechanical gutter cleaners for piggeries and cowsheds are considered below.

Grinding and Mixing Feedingstuffs

Feed grinding was one of the first farm jobs to be made fully automatic. As mains electricity became generally available, it became necessary to develop mills which could be operated by small electric motors, and the result was the development of small automatic mills which will operate either by time switch at off-peak periods and/or will switch themselves off when the required quantity of grain has been ground. This type of mill is so cheap to install and so efficient to operate if well installed that it is unnecessary to argue the case for choosing this general type of equipment. Where the job is too big for one 5 h.p. automatic mill to do it, it is worth while to consider adding a small crusher which can also be set up for automatic operation, or a second identical 5 h.p. mill, or both. There are few farms with needs that cannot be satisfied by an installation comprising a pair of small automatic mills and a crusher. An essential feature of a plant installed for automatic operation is that the meal produced should be delivered either to meal bins or directly to the mixer.

The objects in choosing and installing equipment should be to keep down the labour requirements and so to plan the complete installation that maximum power demand is kept down to a reasonable level. The fact that food preparing can be a wet-day job is not a good reason for unnecessarily wasting labour on it. Where the need for food preparation is a constant one which regularly absorbs several man hours each week, consideration should be given to the possibility of devising a properly planned installation, with bulk storage for incoming grain, a gravity feed to hammer mill and crusher, delivery of the meal to bulk-storage meal bins, or direct to a mixer for the preparation of suitable rations. With a fully efficient mechanical mixing layout, labour requirement need not much exceed 1 man hour per ton mixed, even where bagging is necessary. Mechanical mixing makes home prepared rations a reasonable proposition on an efficiently run holding, so if milling and mixing are considered together a complete small automatic installation incorporating a mixer can be justified for quantities considerably below 50 tons per year.

For example, consider a combined installation incorporating a 3 h.p. 3-phase mill costing £800. Operating costs at various levels of annual usage may be estimated as shown in Table 38, assuming 10 hours use per ton.

Table 38. Costs of milling and mixing with a 3 h.p. combined unit costing £800

	Annual throughput (tons)				
	10	20	50	75	100
Annual hours of use	100	200	500	750	1000
Annual fixed costs £					
depreciation*	67	80	100	100	100
interest	32	32	32	32	32
Total annual fixed cost £	99	112	132	132	132
Annual running costs £					
repairs and renewals †	16	24	36	46	52
electricity ‡	3	6	14	22	29
labour §	8	16	40	60	80
Annual cost £	126	158	222	260	293
Cost per ton £	12·6	7·9	4·4	3·5	2·9

* From Table A.2(i).
† From Table A.3.
‡ Assuming 24 units per ton @ 1·2p.
§ Assuming efficient installation requiring 1 man hour per ton.

Many of the costs are open to argument, but farmers using this basis can substitute the costs that seem to apply more nearly to particular circumstances. Clearly, where quantities are small, even inexpensive equipment will result in fairly heavy depreciation charges per ton, and the investment will not be worth while unless the installation is economical in labour and convenient to operate.

Not all manufacturers use quite the same basis when quoting the capacities of mixing machines. It is advisable to reckon on not more than 28 lb of meal per cubic foot (450 kg/m^3) of the machine's capacity if the machine is to work well when it is filled. As a general rule, a $\frac{1}{2}$ ton capacity mixer is suitable for most medium sized farms, and it costs little more than the $\frac{1}{4}$ ton size. Modern bottom-feed mixers usually employ a high-speed auger, and this has a slight tendency to break up and separate out the cubes in a mixture of cubes and meal.

The requirements of a feed mixing installation vary widely according to the volume to be handled and the variety of rations to be prepared. Some of the governing principles are: (1) Allow for the home-ground meal to be collected in bulk in a closed container, as near as possible to the mill. Blowing meal long distances wastes power and makes it difficult to maintain a dust-free atmosphere. (2) Mixing should be as near as possible to the meal bin, if indeed the mixer itself is not the container; but a working space must be allowed if man-handling of the meal is necessary. (3) Bought-in feed should be kept handy to the mixer, and it may be a considerable advantage to have a reasonably large storage space for such bought-in feed. (4) There should be a one-way flow of meal so that the mixed product is handy for delivery. The ideal flow is in a roughly straight and short line, bulk grain store → grain hopper → mill → meal bin → mixer (meeting bought-in feed) → delivery. It is not always easy to keep the line either straight or short with a complex plant, especially where it has to be housed in an existing building. (5) Where practicable and advantageous, final delivery can well be at lorry height. It may sometimes be an advantage to be able to sack off at either floor or platform level, and this point should be considered where large quantities have to be handled. (6) All conveyors and chutes should be designed in such a way as to avoid creating clouds of dust. The secret of a dust-free atmosphere lies more in avoiding making a dust than in clearing it by means of fans after it has been needlessly created.

In a small modern automatic plant it is often easy and efficient to deliver the main cereal meal straight from mill to mixer, the amount of grain being measured by volume in the grain hopper. The volume method is accurate enough for all practical purposes, and the smaller constituents can be weighed if necessary – though many farmers find it convenient to add a pre-mixed balancer which can be bought in bags of a handy size. This practice of using a commercial cereal balancer makes mixing easier, and is worth considering. The constituents that must be evenly mixed are all incorporated in the pre-mixed balancer, and a good final result is more certain.

With large installations and complex rations arrangements must usually be made for weighing the meal drawn from several different meal bins. A dial weighing machine carrying a tipping container and itself carried on a trolley which runs on a pair of rails from bin to bin and then to the mixer is the kind of solution needed where the quantity to be handled is 5 tons a day or more. For intermediate quantities a fixed weigher near the mixer, and a box or barrel that is easily transported on a pair of wheels may be best.

One method of milling and mixing provides for up to four components to be continuously mixed without the necessity for a batch type mixer, or any manual work. All of the main constituents of the ration pass through the mill, via a multiple-auger feeder arrangement, each channel of which is independently controllable. Another method of achieving the same object is to put a fluted roller metering device on each of the feed bins, and deliver the material on to a single chain and flight or other conveyor which feeds the mill. Where a compact 'mix-mill' machine is used the usual arrangement is to select the quantities from each channel by manually setting dials which control the feed rate of each channel. Very small quantities of additives may be pre-mixed and fed through a separate channel, preferably one which by-passes the mill, so as to avoid heating, and the possibility of destroying vitamins.

On farms where adjacent production units are closely integrated with milling and mixing, the ration may be mechanically conveyed to hoppers in the stock buildings as grinding takes place.

Farmers in many areas can achieve some of the advantages of home milling and mixing, without incurring capital expenditure on milling and mixing equipment, by utilizing the services of contractors who operate mobile mill–mix equipment. The contractor's equipment typically provides for both grinding and crushing, and for mixing the product with any necessary bought materials. The contractor normally supplies the extras, and the margins on this part of the business can help to make the whole efficient. Some of the more expensive mobile outfits are capable of doing work such as grinding straw, which normally is beyond the capabilities of farm plants. Contract prices for straightforward grinding and mixing of cereals may be compared with the range shown in Table 38.

Cubing saves waste in feeding some types of stock, but it is a fairly expensive process and not one that can be recommended to most farmers who make up their own rations. Most of the cost is due to depreciation, repair and maintenance of the machine, rather than to power, labour or other direct operating costs. There are considerable advantages in feeding pellets where spillage is high, as when pigs are fed from dry feeders. Experiments show that pigs fed on suitable pellets produce better live weight gains from an appreciably lower quantity of the same food fed as a meal. Some of the many advantages of liquid feeding systems are discussed on p. 263.

Bulk delivery of meal or cubes can be worth while where large quantities are regularly consumed by individual compact stock units. Where grain and home-produced meals are fed in large quantity in widely scattered units, self-emptying bulk transport trailers which are also mixers, and actually mix the ration during transport may be most economic.

Livestock Feeding

Surveys carried out before mechanized feeding systems were in general use showed a very wide range in levels of labour use for feeding all kinds of livestock. As was to be expected, enormous labour savings were possible where mechanized feeding systems could be introduced in large-scale enterprises; but there was also scope on many farms for labour saving by adoption of simple methods that avoid all unnecessary movement of materials. Hand carrying of bulky fodder is inevitably very time-consuming, especially where the way to the feed points is impeded by gates or other obstructions. Stacking of material such as hay at the point of use can approximately halve the labour need compared with a system that requires use of a trailer.

An A.D.A.S. study of silage-feeding systems showed that a typical labour requirement for hand-feeding of 100 cattle where 60 lb (27 kg) per head has to be cut from a silo and loaded into a trailer is about 3 man hours. The work is cut to about $\frac{3}{4}$ man hour when a hydraulic loader and a suitably modified manure spreader is used for distribution, and to about $\frac{1}{2}$ man hour when a special forage box is used. Further savings are possible with tower silos and press-button feeding systems. Thus, the labour saved by mechanized feeding of a large herd can be an important consideration. It should also be borne in mind that hand-feeding of silage is often a disagreeable job, and generally one which involves very severe physical effort.

Good self-feeding systems exploit to the fullest possible extent the principle of avoidance of unnecessary handling: a good system can compare well with mechanized feeding from the viewpoint of labour need.

In pig feeding, the savings from mechanized systems are generally less than those from mechanized forage feeding because of the smaller quantity of material to be handled and the easier handling characteristics; but labour saving is by no means negligible. An A.D.A.S. study of liquid pig-feeding systems showed that when using a continuous mixer with auto-feed valves, 1000 pigs could be fed and inspected in about $\frac{1}{4}$ hour. Batch-type liquid feeders with hand-controlled valves typically took about $\frac{3}{4}$ hour per 1000 head. Both solid and liquid mechanized pig-feeding systems are commonplace and the main question is not whether to feed mechanically but which of the many variations of both liquid and solid feeding to choose. Similarly, mechanized feeding of dairy concentrates is fully accepted: the main questions are the extent of the need for individual rationing and the extent to which individual variations should be catered for by the use of automatic devices.

Layouts for Self-Feeding Silage and 'Easy-Feeding'

Successful self-feeding of silage calls for careful planning to ensure that labour saved in the feeding operations is not needed for extra cleaning and littering. The principles have quickly become well established, and the wide variations in practice are necessitated by different sets of buildings, and the need to feed different amounts of silage.

The most usual need is to self-feed *ad lib.*, the cattle being kept in covered or partly covered straw yards or in cubicles. With such an arrangement the following points are worth notice:

(1) In straw yards, the resting area should be quite separate from the feeding area, though it may be beneath the same roof. With de-horned or polled cattle, as little as 45 sq ft (4·2 m²) of resting area per beast can suffice in good conditions.

(2) With 24-hour access to the feeding face, 9 in (23 cm) per head is ample. One or two small lights facilitate night feeding.

(3) Place the silage on a concrete apron in such a way that slurry tends to flow away from the silage at the feeding face.

(4) Do not attempt to bed down the concrete at the feeding face. Keep the area as small as possible and arrange for regular cleaning.

(5) Have a small unbedded concrete walk between the feeding area and the resting area, so that most of the slurry is dropped from the cows' feet before they enter the resting area. If necessary, the cows can be made to walk a few extra yards by the use of an electric fence. This helps to keep the bedding clean.

Similar considerations apply to almost any self-feeding or assisted-feeding layout.

There are very many variations of layout which have produced successful results. One of the most common is to house the silage (with straw above) in the main span of a Dutch barn, and have the resting area in a comfortable lean-to. Movement from one to the other can be by way of a small open yard placed across one end of the building. For larger numbers of stock, a longer barn can be fitted with a lean-to at both sides and open yards at both ends, so that two quite separate herds can be self-fed under the same roof. If rain is completely excluded from the bedded area, care is taken with the lay-out, and regular cleaning is practised, as little as $\frac{1}{2}$ ton straw per head per winter should suffice, whereas with bad plan-ning, involving a bedded area immediately adjacent to the feeding face, the amount of straw needed will be more than doubled.

With any self-feeding system, uniform eating of the face depends more on a uniform quality of silage than it does on the design of feeding barriers. It is, however, worth noting that a good deal of unnecessary waste can be prevented by making the stock keep their fore feet well clear of the silage face. This may be done by the use of a rounded pole, of about 6 in (15 cm) diameter, which is simply laid on the ground a little nearer to the cattle than the electric fence.

The chief difference between assisted rationed feeding and *ad lib.* self-feeding lies in the need with rationed feeding to provide a very long feeding face, preferably with a rigid division for each animal. This makes 'assisted self-feeding' usually only suitable for relatively small herds. The method of operation in this case is usually to take away one long side of the silo and to bring up heavy feeding barriers which enable a rationed amount of silage to be cut and forked direct to the stock without the use of any transport. Naturally, the consider-ations regarding cleaning and littering are very similar to those for *ad lib.* self-feeding. The chief disadvantage is that with the long feeding face exposed, a certain amount of moulding of silage is difficult to avoid.

Mechanized silage feeding may be preferable where moder-ate rationed amounts have to be fed. One method is to provide a long feeding rail with a place for each animal, and to cart the silage to the feeding platform by means of a specially modified silage fork mounted on a front-loader. The silage is cut in vertical layers, and loads of $\frac{1}{4}$ ton or more are easily handled. For transport over longer distances 2–3 such loads can be put on a buckrake. With careful handling of the slabs of silage it is possible to put a week's supply of silage in position near the feeding fence, and the stock-man then handles the silage from the heaps by hand fork. No serious moulding has been found to occur with the system in normal weather during the winter and early spring months.

Mechanized Cattle-Feeding Systems

Mechanized cattle-feeding systems may be divided into two main types: those employing a mobile forage box for feed distribution and those with fixed 'press-button' conveyors. The latter are usually linked to tower silos for conserving high-dry-matter silage, or high-moisture barley, or both. The tower silo provides a good basis for systems engineering, since it confines the silo unloader within its circumference, and makes it comparatively easy to control it, either from a fixed axis in the centre of the silo or outside the circumference, in the case of bottom unloaders, or by means of a triangular framework with wall-riding wheels in the case of top unloaders. Material such as baled hay, on the contrary, does not lend itself to highly mechanized feeding systems. With bales, even if final delivery of the fodder to the cattle is by a mechanical feeder, each individual bale has first to be fed by hand on to a conveyor. If the general adoption of hay wafering becomes possible, this would revolutionize prospects for mechanical feeding, and extra production costs similar to those in California, of £3 to £4 per ton above the cost of baled hay, could easily be justified. Barley is much easier to handle than roughages; and short-chopped dry silages are easier than materials which are wet or include long lengths of fibrous materials.

Tower Silo Unloaders

One of the first questions that needs to be settled when deciding details of a tower silage system is what type of unloader will be used. Early top-unloaders all employed either single or double augers for loosening the silage but endless chains with cutters are now available for both top- and bottom-unloaders and there are considerable variations between makes in how these are driven and how the loosened material is conveyed to the feeder. The main types may be listed as follows:

Top-unloaders
   Auger(s) and impeller, using external chute for delivery.
   Auger using central flue formed in the silage.

Rotating arm chain and cutters, with second chain delivering to internal side chute.

Rotating arm chain and cutters with vertical auger and conveyor-belt discharge to external chute.

Bottom-unloaders

Rotating arm with chain and cutters. Delivery by second chain.

Swinging arm with chain and cutters, working from pivot outside silo. No system has all of the advantages and no disadvantages. All types have been greatly improved in performance and durability as a result of continuous development to suit the testing conditions that have to be overcome when handling typical grass silages, which are much more difficult to handle than maize.

Chief disadvantages of the simple auger/impeller type are a fairly high power requirement and the fouling of the access chute and ladder which occurs with all types that employ a dual-purpose chute. With the central chimney system a fairly extensive silage face is exposed to air along the central chute; and this can lead to severe over-heating if over-dry silage is made. The aim should be silage of about 30 per cent dry matter.

A system with many advantages is one which employs a pair of endless chains working on top of the silo, and unloading into a chute which is built just inside the silo doors and is dismantled and reassembled from the top downwards as unloading proceeds. The endless chain system has a high output for a low power requirement. The lined chute ensures a minimum of waste due to secondary fermentation; and the personnel ladder and protective cover can be designed purely for safe access to the silo, and kept free from silage. As with all top-unloaders, it is necessary to go up the tower once every few days to adjust the delivery arrangements.

One of the main advantages of a bottom-unloader is that the climb up the tower to adjust delivery is unnecessary. The other main advantage is that filling of a partly-used tower can take place at any time – even during feeding. A disadvantage with all bottom-unloaders is that considerable disturbance of the lower layers takes place before the silage is removed, owing to

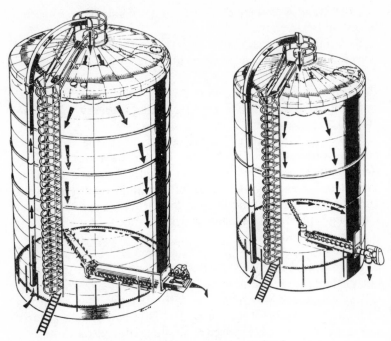

Fig. 10. Bottom-unloaders with sweep-arm mechanism.
Left: Chain type for forage. *Right:* Auger type for high-moisture grain.
(Howard Harvestore)

the way in which the silage falls down from a hollow dome above the unloader. With dry silages, harmful heating can result from air entering via the unloader.

Mechanical Cattle-Feeders

When planning provision of mechanized cattle-feeding equipment, the first consideration should be the requirements of the type of stock concerned. If the job to be done is a simple one, such as feeding silage only to a single dairy herd, any of the main types of equipment listed in Table 39 may be chosen, and the simplest will suffice. If, however, it is desired to feed a line of cattle pens, containing animals of different ages and sizes, with different amounts of fodder; or to feed one ration on one side of the feeder and a different one on the other side, choice is

Table 39. Main Types of Fixed Mechanical Cattle-Feeders

Type of Feeder	Full-length or progressive	Different yards	Pens along Feeder	Power needed per 120 ft (36 m) length or equivalent h.p.
			Rationing facilities	
Open-bottom auger	Progressive	Not suitable	Not suitable	$5-7\frac{1}{2}$
Slotted tube auger	Full-length	Either side by baffle	Adjustment of position of openings, in sections	$5-7\frac{1}{2}$
J-type auger	Full-length	Either side by baffle	Adjustment of position of opening, in sections	3
U-type auger	Full-length after short interval	Either side by baffle	Adjustment of size of openings	3
Three-way auger	Full-length	Either side or far end	Adjustment of gap in sections	$5-7\frac{1}{2}$
Endless belt with scraper or brush	Full-length	Either side	Adjust travel of scraper or brush	
Chain and flight with tapered bed	Full-length	Either side by baffle	Not suitable	$1\frac{1}{2}$
Chain and flight or endless belt 'cascade'	Full-length	Either side by baffle	Not suitable	1
Chain and paddle or belt in manger	Progressive	Not suitable	Not suitable	$1\frac{1}{2}-3$
Chain and flight in circular manger or rotating manger round foot of tower	Progressive	Not suitable	Does not apply	2/3 to 1
Jog-trough	Progressive	Not suitable	Not suitable	

(Normally used only for bulky, light, dry materials, e.g. hay or straw)

more restricted. Though reliable information on repair and maintenance cost is difficult to obtain, owing to limited use of some types of equipment, this factor should not be overlooked.

One of the most adaptable methods of cattle-feeding, and certainly the cheapest in capital cost where cattle of differing requirements are housed in several scattered yards, is by means of a mobile forage box. The latter is usually tractor-operated, but possibilities of putting the feeder box on a truck chassis may be worth exploring in the case of large, scattered installations. In either case, the mobile equipment is at a disadvantage compared with good compact fixed installations when judged on the basis of ease of operation at weekends or in a busy time.

In making a choice, particular attention should be paid to

matching and linking up the feeders with the unloader delivery. Often, expensive intermediate conveyors are necessary, and these can greatly add to both capital cost and maintenance problems. Repair costs can be a particularly serious item in all kinds of feeding equipment if, owing to a very poor unloader output, the equipment has to be run nearly empty for long periods. Since the manufacturer of a feeder which is starved by a slow unloader can justifiably claim that this does not give his equipment a chance to perform well, there is much to be said for making one firm responsible for the whole unloading and feeding installation.

With the common types of feeder, the amount of trough space required for rationed feeding is 24–28 in (60–70 cm) for large, adult cattle; 16–20 in (40–50 cm) for calves and 20–24 in (50–60 cm) for cattle of intermediate size. Much smaller lengths suffice where *ad lib.* feeding is practised, as with intensive beef production.

Some types of feeder, such as progressive open auger feeders, are not suitable for use where grain or concentrate is the main feeding-stuff.

Most feeders can deal with mixtures of concentrate and roughage, but some types tend to sift out a higher proportion of the concentrate at the supply end. An advantage of endless belt feeders, is ability to feed mixtures, without separation, throughout the length of the manger.

Feed Dispensers for Milking Parlours

Though there are arguments against feeding concentrates in the milking parlour, this is normal practice, and is the most convenient arrangement that can be made on most farms, whatever the system of forage feeding employed. As a general rule, however, an effort should be made to devise other feeding arrangements so that the quantity of concentrate to be fed in the parlour is reasonably small. Thus, if it is necessary to feed barley in order to balance a silage that is rich in protein, there is everything to be said for feeding at least a part of the barley with the forage – e.g. sufficient to feed for a yield of up to 3 gallons (14 l).

The common method of parlour feeding is to provide a hopper and dispenser for each stall or pair of stalls. In addition, it is usually necessary or desirable to provide equipment to convey the concentrates to the hoppers, and this may add considerably to the total cost.

A cheaper type of installation which is suitable for a herring-bone parlour employs a moveable-hopper type of dispenser, which runs on tracking above the troughs.

For conveying meal or crumbs of up to $\frac{3}{16}$ in (5 mm) diameter to stationary hoppers, one of the cheapest types of conveyor incorporating a light chain and flight mechanism may be used. In one type, discs on an endless wire are drawn through a tube of about $1\frac{1}{2}$ in (3·7 cm) diameter. Such a conveyor can elevate at any desired angle, and can operate round right-angle bends. The low capacity of about $\frac{1}{2}$ ton per hour is usually quite adequate for parlour feeding, and a fairly long run from the mixing room to the parlour is not ruled out. An alternative light type employs a chain and flight mechanism in a U-shaped trough.

Where a high conveying rate is needed, a 3 in or 4 in (7·5 or 10 cm) diameter enclosed auger may be used, and is suitable for cubes of up to $\frac{3}{8}$ in (9 mm) diameter. Capacity usually ranges from $\frac{1}{2}$ to 2 tons per hour.

There is a wide choice of equipment for reducing the labour and the decision-making in the work of the parlour operator. The primary and essential step is automatic replenishment of the dispensing hoppers from the bulk feed-store, or an easy way of doing this by hand. The next step is to centralize the operator controls and to do the dispensing electrically instead of manually. This usually requires the operator to adjust the amount supplied to each animal by setting a scale for each stall. The decision-making may be simply assisted by attaching a colour-marking tag to each animal, so that the herd is divided into a suitable limited number of classes according to the amount of concentrate needed. The automation process is taken a stage further by a system in which the operator simply records the cow's number, and a small computer is used to control the amount of feed given. Other systems go further, and involve fitting the cow with a small electronic device which provides for automatic cow-recognition, and so removes this job from the operator's routine. By comparison with these, a system which feeds according to yield as the cow is milked is relatively simple.

In comparing simple with more complex systems, a factor that needs to be considered in addition to cost and reliability is that independent experiments indicate that there is little or nothing to be gained by individual feeding compared with rationing on a sensible group basis.

Automatic Calf-Feeding

A common type of automatic calf-feeder consists of a stationary device with a single teat which provides a ready-mixed supply of liquid feed for a group of up to about 15–20 calves housed in a single pen. This is not, however, the only kind of automatic feeding system. In another kind the feeder is mobile, and moves automatically along a line of single-calf pens, so that each individual calf has equal opportunity of feeding, regardless of size or temperament. A third system is mechanized but not automated. The feeding arrangement consists of a parlour with individual stalls and a rationed service from a bulk supply to each stall. The calves are driven to the parlour in batches of 20–30, and when they have finished, the feeding equipment is automatically washed ready for the next batch before the new supply is delivered.

Stationary Automatic Feeder for Group of Calves

The feeder usually consists of a hopper which is filled daily with a suitable milk substitute powder; a supply of water and electricity; and the necessary thermostat, heater and mixer to provide a ready-mixed feed at blood heat whenever the supply is removed. In a typical cycle, a calf drinks the quantity mixed, which is 1–2 pints (0·5–1 l); this automatically cuts off the teat from the supply, and starts the mixing of a standard quantity of powder in a selected quantity of water. Mixing is completed in about 3 minutes, and the newly mixed feed then becomes available. Provided that the feeder is not overloaded by having too many calves for the single teat, it does not matter if an individual calf takes two or three successive feeds, or is satisfied with the 'little and often' routine, to which most calves soon settle down.

Economics of Mechanized Cattle-Feeding

Mechanized feeding of forage from tower silos has been briefly discussed in Chapter 7. It is not difficult to justify the cost of fully mechanized feeding on a dairy farm which is able satisfactorily to increase the number of cows kept on a given acreage, without adding to labour costs. There is evidence from commercial farms that mechanized feeding, if well planned and operated, can result in higher all-round efficiency in milk production, due principally to less waste of food and more efficient use of labour. On some farms where records have been kept, results of a partial change to mechanized feeding have been good enough to persuade the farmer to purchase more silo capacity and equipment in order to complete the switch. It is not practicable at present to lay down standards of levels of capital investment which can be justified. A completely new installation of cattle yards, tower silos, mechanical feeder, milking parlour and manure disposal arrangements for a herd of about 120 is likely to cost in the region of £30 000 to £40 000 or £250–£350 per cow. Even at £250 per cow, annual cost, reckoned on a 10-year depreciation basis, at 15 per cent per annum including maintenance and interest, amounts to £37·5 per annum per cow. If it is possible to average 1000 gallons per head, the fixed costs of the equipment amount to about 3·75p per gallon (0·8p/l). It is therefore not surprising that completely new installations are rare. The cost of a mechanized feeding system involving tower silos depends on the type and size of individual silos, the length of the silage-feeding period, feeding policy and type of stock. As a rough guide, Table 40 shows the silo capacity needed for dairy cows, reckoned on the basis of $1\frac{1}{4}$ cu ft (35 dm^3) of silage per head per day. Beef cattle should usually be allowed about $\frac{1}{2}$ to $\frac{3}{4}$ cu ft (14–21 dm^3) per head per day, and sheep $\frac{1}{4}$ cu ft (7 dm^3).

Efficient silage-making requires the silo to be filled at a minimum rate of about 10 ft (3 m) per day. It is necessary to avoid use of silos of excessively large diameter in relation to the

Table 40. Tower Silo Capacity Required for Dairy Herd,
Reckoned at $1\frac{1}{4}$ cu ft (35 dm^3) Per Head Per Day

| Size of herd | Length of Feeding Period (days) | | | | | |
| | 120 | | 180 | | 240 | |
	cu ft	m^3	cu ft	m^3	cu ft	m^3
60	9 000	250	13 500	380	18 000	510
70	10 500	300	15 750	450	21 000	600
80	12 000	340	18 000	510	24 000	680
90	13 500	380	20 250	570	27 000	760
100	15 000	420	22 500	640	30 000	850
120	18 000	510	27 000	760	36 000	1 020
140	21 000	600	31 500	890	42 000	1 190
160	24 000	680	36 000	1 020	48 000	1 360
180	27 000	760	40 500	1 150	54 000	1 530
200	30 000	850	45 000	1 280	60 000	1 700

possible daily work rate of the field and transport equipment.
Table 41 shows approximate work rates required for different
sizes assuming that all are about 65 ft (20 m) to eaves and
allow 60 ft (18·3 m) settled depth of silage.

Complete filling is never practicable, and actual capacity
depends on the extent to which the silo is re-filled after settling.

It should be noted that the weight of silage is not directly

Table 41. Work Rates Related to Silo Sizes

| Silo diameter | | Floor area | | Approx. dry matter stored at settled depth 60 ft (18·3 m) ton | Approx. total weight at 40% d.m. to be ensiled per 8-hour day ton | Approx. work rate needed ton/hr |
ft	m	sq ft	m^2			
18	5·5	255	23·5	138	53	7
20	6·1	314	29·0	169	65	8
22	6·7	380	35·3	204	78	10
24	7·3	455	42·3	243	94	12
26	7·9	530	49·0	286	110	14
28	8·5	618	57·2	331	127	16

related to height of the silo, but tends to become greater per unit of volume with increasing height.

Where suitable yards and a parlour already exist, the addition of tower silos and mechanized feeding for 60, 120 and 180 cows is likely to result in capital costs on specialized equipment approximately as follows:

Capital cost of mechanized feeding installations for dairy herds (£)

	Size of dairy herd (6 months feed)		
	60 cows	120 cows	180 cows
Silos, including base, erection and filling pipes	3 500–5 000	5 200–8 700	8 700–11 500
Unloader	1 000–2 000	1 100–2 100	1 100–2 100
Dump box	1 200–1 300	1 200–1 300	1 200–1 300
Blower	400–500	400–500	400–500
Fixed feeder	600–900	1 000–1 400	1 400–2 000
Electrical wiring and other on-site installation work	500–600	550–650	600–700
Total installed cost £	7 200–10 300	9 450–14 650	13 400–18 100
Cost per cow £	120–170	79–122	74–100
Annual cost per cow based on 10 year life at 15% p.a. £	18–25	12–18	11–15

The above figures are intended only for illustration. The reader can substitute figures for specific items of equipment, and for any additional or alternative costs related to particular circumstances. A considerable amount of additional field equipment is needed for any efficient method of fodder conservation, and this must be included if total costs of making and feeding silage are sought.

In considering whether such capital expenditure can be justified, it is necessary to take account of many factors besides the not inconsiderable saving of labour. Surveys of farms before and after installation of towers show that when a sound system is introduced by a competent silage-maker there can be either an appreciable increase in herd size, or an appreciable area of land saved for growing another crop. The extent of such savings naturally depends on the starting point but can be as much as 20 per cent. With dairy herds an appreciable saving in the cost of purchased concentrates can be expected from more efficient conservation and earlier cutting. It can be argued that some of these improvements could be achieved with less expensive systems; but the fact is that the stimulus of an improved system is often the key to a general improvement in grassland management.

Where the cattle enterprise is beef production it has not been easy to justify high capital expenditure on mechanized feeding, because the cost of young cattle, and prices received for fat cattle are variable; and average returns can only pay for the installation if the equipment can profitably be used year in and year out at fairly full capacity. On individual farms this usually implies an installation for a large number of cattle, and a total size of enterprise which is too demanding in capital for many farmers. It should be noted that beef production has to compete indirectly with low-cost systems of production, such as extensive ranch systems in Argentina, and very large 'feed lots' in some parts of U.S.A. For example, typical feed lots in California need only very crude shelter and have few problems in manure handling, owing to the hot, dry climate. The main feed is usually grain rather than forage; and where forage is fed in large feed lots it is usually maize stored in very large bunker silos, with mechanized unloading and forage-box feeding. The minimum size for really low costs is reckoned at about 4000 head; and on this basis, capital cost can be kept down to something in the region of £10 per head.

In Britain the main objectives connected with beef enterprises are varied. On a large corn-growing farm the main function of the beef enterprise may be to make good use of the leys which are considered essential to the rotation. On another farm the main objective may be quite different, e.g. to increase output from the farm by buying in barley to feed. In either case, the capital costs will be considerably reduced if suitable covered yards are already available. A mechanized feeding installation designed to handle both forage and high-moisture barley is often the most desirable; but it unfortunately costs more than one for feeding either forage or barley only. Storage facilities for a typical installation for 300 head fed for about 200 days may consist of two forage silos about 20 ft (6 m) diameter by 60 ft (18 m) high, and one high-moisture barley tower of about 250 tons capacity. Capital cost of the towers, silo filling and emptying equipment, food preparation and mechanical feeding equipment is likely to be about £20 000 or £67 per head. Where a new barn has to be erected and new arrangements made for manure handling, the cost will, of course, be considerably higher. Such capital costs are not prohibitive in themselves. At a capital cost of £67 per head, the annual charge including maintenance and interest at 10 years' depreciation amounts to about £10 per head. If provision of an elaborate special-purpose building results in double this capital cost, the likelihood of a substantial profit is remote. One of the difficulties in beef production is the high capital investment in the stock. Even if calves are bought and reared, the initial outlay on 300 head is likely to be about £10 000. Interest on capital and availability of capital are therefore factors of high importance to the success of such an enterprise.

Mechanized Pig-Feeding

There are many successful systems of mechanized pig-feeding, and the advantages of mechanized feeding are sufficient to make it virtually certain that hand methods will quite soon be found only where pigs are not a major part of the farming enterprise. General considerations which should be borne in mind in choosing a system include the following:

(i) The cost of labour is small compared with food cost. With an un-mechanized system it is typically not more than 10 per cent, compared with food cost at about 75 per cent. Moreover, a high proportion of the labour may be devoted to other parts of the job, such as manure removal. It follows that there is less point in a highly mechanized feed system if the real labour problem lies elsewhere.

(ii) Though *ad lib.* feeding is suitable for some purposes, such as the production of heavy hogs, and though it is possible to devise rations which are well suited to such a feeding method, the usual requirement is rationed feeding at regular intervals. One of the advantages of *ad lib.*

feeding is that it is possible to make do with much less trough space per pig. For rationed feeding 50 lb (23 kg) pigs need about 8 in (20 cm) per pig, and 150 lb (68 kg) ones need 12 in (30 cm).

Feeding Systems

In the simplest systems of *ad lib.* feeding, a horizontal auger or chain-and-flight conveyor fills the hoppers of meal or pellet feeders by means of telescopic down-spouts which dip into the hoppers and automatically cut off the supply at a pre-determined level. The hoppers are progressively filled, and the conveyor is automatically switched off by a pressure switch when the farthest hopper is full. If such an arrangement is installed in a new building it makes it possible to dispense with a feed passage if desired, an inspection walk-way being provided above the pens.

There are several systems of rationed feeding of dry meal. One of the earliest to be fully automated incorporates a travelling meal hopper, which has its metering device driven by a series of pegs above the troughs. As the hopper passes above the trough, each peg causes a given amount of meal to be dropped, the number of measures delivered to individual troughs being determined by the number of pegs that are in place. This equipment is winch-operated, and was designed primarily for trough feeding with a central feeding passage.

For on-floor feeding the most common systems employ augers or endless chain conveyors to fill metering hoppers with hinged bottoms. The amount of food held by the hopper is easily adjusted, and at feeding time the release mechanism is usually tripped manually. This system can, however, be fully automated, so that the only attention needed is filling of the bulk supply hopper, adjustment of the metering hoppers as necessary, and, of course, inspection of the stock. When feeding is automated, arrangements are usually made to cause a bell to be rung just before feeding begins in any house. This enables the stockman to go to the house and inspect the pigs as they feed. Automation, by freeing the operator from chores, certainly makes it easier for him to make the necessary regular checks on health and thriftiness of the stock.

Wet-feeding systems for pigs have many advantages. The

necessity to mix the meal and water before pumping it to the troughs makes it practicable to do without any other form of mixer. Home-ground meal and bought-in additives are typically fed into a metering hopper in the desired amounts to make up the mix. Typical mixer sizes are 400–800 gallons (1800–3600 l). The mixer is partially filled with the required amount of water; the stirrer, usually a 9 in (23 cm) diameter auger, is started; and the meal is gradually added. After about 10 minutes the mix is ready for pumping. Typical power requirements are 3–5 h.p. on the mixer and 5–10 h.p. on the pump. Usual method of distribution is by hand-operated taps above the troughs, sometimes assisted by a short length of flexible hose to enable two or more troughs to be served by one tap. Polythene pipes are suitable, typical sizes being 2 in diameter for general distribution and $1\frac{1}{2}$ in for any flexible outlets. A complete loop is arranged, so that surplus feed returns to the mixer, and there are no lengths of pipe in which meal remains after pumping.

The normal strength of mix is about 4 lb meal to 1 gallon of water. Rationing may be by time, or by marking levels in the trough.

The machinery for wet feeding is simple, and the absence of dust a definite benefit. There is hardly any practical limit to size of the installation. For example, there are small units which are reasonable in cost for a building housing only 200; and it is quite possible for one man to feed some 5000 pigs from a single large unit, provided meal handling is fully mechanized and mixing arrangements are partially automated. With continuous mixing and use of automatic valves the amount of time needed for feeding is simply that required to pre-set the controls and to inspect the pigs as the food is delivered in succession to each pen according to need. With either automatic controls or hand-operated valves, rationed feeding is practicable; and experience shows that the liquid system can be successfully applied to bacon production as well as to the production of heavy hogs. Results of experiments at Experimental Husbandry Farms indicate slightly higher live-weight gains and a slightly better conversion ratio for liquid feeding compared with dry, but this is probably associated mainly with avoidance of dust and of waste.

Capital Costs of Mechanized Pig-feeding

Capital costs of mechanized feeding systems can be satisfactorily low provided that in the case of the more elaborate systems the cost is spread over a reasonably large number of pigs. For example, a fully automated liquid feeding system for 3000 pigs can be installed for the low price of about £2 per pig place. It is naturally difficult to secure low capital costs with small installations; and for 200 pigs the cost of many systems is likely to exceed £5 per pig place. Generally, however, it should be possible to find a suitable system for 200 pigs or more for which the capital costs can be easily justified by savings in time, ease of doing the work and considerable potential advantages of less waste and a better conversion ratio which are possible with a well engineered system.

One of the advantages of a good mechanized system is that the stock become so accustomed to it that they take little notice of anything but operation of the feed mechanism, and so rest a great deal more than when hand-fed. Farmers who have kept records over a period covering a change from hand to mechanized feeding have generally found improved results as soon as the pigs settle down to the new system. There can be little doubt that mechanized feeding will soon become normal practice on all but the smallest farms.

Use of Electric Fences

Electric fences permit the folding of fodder crops, protection of ricks, control of cattle in yards, self-feeding of silage and the driving of stock along unfenced roads and past arable crops. An official bulletin gives full details concerning all aspects of construction, maintenance and use. Nothing need be said here concerning the construction of fence controllers, and little concerning types of posts and insulators. For normal folding purposes there is much to be said for simple polythene insulators which are unbreakable, and from which the wire can easily be disconnected when moving the fence. Very light pointed steel fencing posts specially designed for the job are usually handiest for folding, but there are occasions when the ground is very hard when such ideas as fixing the posts to old motor tyres or to blocks of concrete are worth considering.

The first point to observe about using fences effectively is that it is worth while to take care to introduce stock to the fence in a manner which enables them to be quite sure that it is touching the fence – however gently – that gives the unpleasant shock. This must be done when the animals are quiet and unafraid. The worst possible method of introducing stock to the fence is to drive them on to it, since they will probably run straight through it and will still not understand what caused the pain. One method of training stock is to put them quietly into a small specially erected training fence and leave them to reach for tempting food placed just beyond the fence. For general purposes, the number of wires and heights required for the common types of stock are as follows:

Adult cattle – One wire at 30 in–42 in (76–100 cm) according to size.
Calves   –   ,,   ,,   ,, 18 in–24 in (45–60 cm)
Adult pigs   –   ,,   ,,   ,, 12 in–15 in (30–40 cm)
    (Not really suitable for sows with litters.)
Sheep   – Two wires at 12 in–15 in (30–40 cm) and 24 in–27 in (60–
                70 cm)

Chief use of the fences is for getting the best out of pastures that are grazed by cattle. Here a choice has to be made between rotational grazing and strip folding. For dairy cows, strip folding is usually economically justified, and it is worth while to divide up large fields into blocks of a convenient size which can be strip folded in turn. (Fig. 11.) The semi-permanent electrified fences remain in position for most of the grazing season, and with such an arrangement either strip folding or rotational grazing can be carried out in each of the paddocks in turn, according to need. The service paddock can be eliminated if necessary, but though there are disadvantages in the constant grazing that occurs there (e.g. no break in cycle of sward-borne internal parasites) this system is often best on balance.

When an electric fence is used for self-feed silage, a suitable method is to suspend a galvanized steel water pipe in front of the silage face, and connect this to the electric fence. The pipe is, of course, suspended on insulators.

An electric fence can also be used effectively for bringing

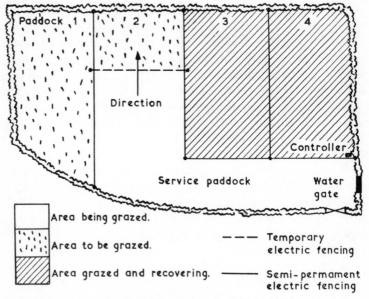

Fig. 11. A method of strip folding using an electric fence.
Source: Min. Agric. Bulletin 147 *Electric Fences.*

cows forward towards the entrance to a milking parlour. The so-called 'electric dog' consists of an electrified bar which is suspended across a pair of parallel wires at the far end of the yard. As milking proceeds the operator brings any reluctant animals forward by simply pulling on a rope which moves the electrified bar nearer to the parlour entrance.

Use of single-strand galvanized wire is a waste of money, due to stretching, kinking and general difficulty in handling. Multi-strand steel wire is more flexible and has a high tensile strength. Some manufacturers supply a multiple strand plastic thread with the metal conductor wires interwoven.

Cleaning and Littering

Some information on handling and distribution of manures has already been given in Chapter 6. Where straw yards are used for cattle, littering is a routine job which may take a considerable amount of time. While a fully bedded and completely covered beef yard may need only 7 lb (3 kg) of straw per head daily, a cow yard will typically need at least 12 lb (5 kg) and often much more. Distributing the straw by hand can be a time-consuming job, and it is worth while to consider using a flail type trailer manure spreader to distribute the bales, after the strings have been removed from a layer placed, one bale deep, in the bed of the spreader. This necessitates designing the yards so that the spreader can drive through easily. The system is also suitable for cubicle yards where the manure is scraped and handled in solid form.

Where a herd of about 60 cows lives in semi-covered yards, tractor scraping of the concrete is usually a regular chore, which even if done only 3 times a week will often take 6 to 10 hours of man and tractor work weekly for scraping and distribution. At such a cost, the manure needs to be well utilized if the fertilizer value is to pay for the cost of handling. Many types of tractor-mounted yard scrapers are now available, and are well worth their small cost.

Where cow-sheds are used, the usual methods of hand cleaning cannot be carried out in much less than one hour daily for a 60-cow herd. Most of this time can be saved by use of a mechanical gutter cleaner, which is likely to cost about £1000 to install. Such equipment is well worth considering where the design and situation of the building are such that a tractor scraper cannot be effectively used. A tractor scraper is considerably cheaper in capital cost if the tractor itself is available for the work. Any mechanized method is likely to require some re-shaping of gutters, etc. to suit it. Though the direct effect of mechanization is labour saving, the usual justification for capital expenditure is not saving a man but being able to keep more cows with the staff available.

An ideal manure handling system is one which needs no attention except for the actual distribution on the land. Many piggeries have fully automatic systems which do not require manual work or the use of mechanical equipment, the manure which passes through gratings in the dunging passage flowing over correctly designed weirs to the storage tank. Other systems for piggeries involve the use of mechanical scrapers. Most require some supervision, but can save a considerable amount of labour. The main effect of introducing a system which requires little labour is that it permits many more pigs to be kept, or more time to be given to other productive work. Mechanical equipment can be installed for about £2 to £5 per pig if provided when the piggery is constructed. This can save much of the average of about 5 man hours per pig per year expended on cleaning out piggeries where the job has to be done entirely by hand scraping.

Many types of poultry laying cages incorporate fully automatic manure removal by means of a slowly moving endless belt, or a chain and slat conveyor. Such equipment is rapidly becoming essential for low-cost egg production.

Milk Production and Milking

Developments in mechanization and automation rapidly make obsolete any figures for labour use on various aspects of milk production, whether the figures are in percentages of a man's time or in terms such as man minutes per cow. For example, average work hours per dairy cow per year were 150–180 in the 1940s, and are now less than half of this. Variations between individual farms are very large. Great reductions in labour use were made possible on some farms by changing over from cowsheds with bucket milking plants to yard and parlour systems. Bulk handling of the milk and the use of advanced types of milking parlour added to the savings. The effect of this is that whereas 30 cows milked per man was not long ago regarded as a good target and is still fairly common with small herds kept in cowsheds, 60 cows per man has become commonplace even on fairly small farms, and 80–100 not very rare on well-equipped large farms, where other workers contribute to feeding and relief work. The savings in cost of milk production arising directly from labour saving are significant; but the average rate of progress is inevitably slowed by the existence of many small farms and small herds, a good proportion of which find it impracticable to change from keeping the cows in cowsheds.

Though milking constitutes the biggest labour item on most dairy farms, averaging about 40 per cent of the total, it is sometimes possible to make important savings in other items, notably on forage feeding and on cleaning, which typically account for about 20 per cent each of the total work load. By cutting the time needed for these jobs, more time can be found for milking and this may make it possible for more cows to be handled. For example, a feeding job for a herd of 60 which takes 3 hours by use of a hand-loaded tractor trailer may take only $1\frac{1}{2}$ hours by an 'easy-feed' system where the handling distance is negligible, and may be cut to $\frac{1}{2}$ hour using a hydraulic loader and forage box, or to $\frac{1}{4}$ hour by a tower feeding system with fixed conveyors.

Scraping slurry or hand cleaning of cowsheds are regular chores which take up to an hour daily on many farms. It may be possible to save most of this time by a well-planned layout of passages suited to tractor scraping, or by use of slatted passages in a cubicle house. Such advantages may tip the scale when deciding whether to spend money on modernizing or replacing unsuitable buildings. Similarly, some farmers who long ago changed over to straw yards want to be rid of the time-consuming job of distributing straw litter, a task which can take on average almost an hour daily. The fact that less time is needed to distribute straw or sawdust in cubicles will certainly be a consideration in deciding whether to change over to a cubicle system. With a good layout of cubicles straw consumption can be as little as 0·15 tons per cow per winter.

Milking

The time taken for milking steadily falls as machines are improved in detail, and are designed to do more and more of the work which was formerly done by hand. A study in the Eastern Counties in the late 1940s, at a time when hand milking was still fairly common, showed that labour on average represented about one-third of the whole cost of milk production, and that milking accounted for 40 to 55 per cent of the total. At that time the potential for mechanization was well indicated by the fact that the 10 farms with the lowest annual labour costs averaged 93 man hours per cow, compared with 309 for the 10 farms using most labour.

Since that time, great progress has been made on some farms

in reducing the labour requirements for milk production as a whole, and especially in respect of labour used for milking. With a large herd in a well-planned herringbone parlour the work routine has been reduced to as little as 0·8 minutes per cow. This means that where, as in a large herringbone parlour, enough units and stalls are available, it becomes possible for one man to milk up to 70 cows per hour. In practice, such high rates are seldom yet achieved; but already, the best-equipped and managed farms have working rates much better than the average of 23 cows per man hour for herringbone parlours found in a 1964 survey. Methods of cleaning parlour equipment can now be almost completely automated, and results in terms of hygienic milk production can be more satisfactory than where a considerable amount of time is employed on older methods.

Choice of Milking Parlour

Milking routines have been studied in detail by many workers, and the results of these studies enable positive conclusions to be drawn covering most types of installations that are of interest today. Choice of a parlour to suit a particular farm involves consideration of many factors, and it is worth while to study the main questions that must be considered.

In the space of a few years it has become generally agreed that pipe-line or direct-to-churn milking is ideally a one-man job, and it is as such that it is largely considered here, though there are parlours such as the herringbone type which are designed to allow two men to work in one parlour more or less independently.

Types of Parlour

In the abreast type of parlour the cows stand side-by-side, as in a cowshed, and the operator works behind and between them. He may have a milking unit to each stall or one unit between each pair of stalls. Because of the need to have a level floor for the operator to work on, and the fact that cows cannot be expected to step up more than about 14 in (36 cm) this is about the limit of the difference in floor levels for abreast parlours. In a tandem parlour the cows stand in a line, head-to-tail, and the operator works at a lower level alongside them. The line of cows is usually straight, but need not be so. Single-sided parlours may be L-shaped or U-shaped (Fig. 12). In double-tandem

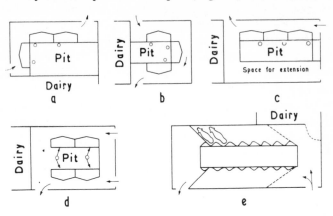

Fig. 12. Some layouts of two-level milking parlours.
a. L-shaped, *b.* U-shaped and *c.* in-line tandem parlours with one stall per unit. *d.* 2-unit, 4-stall tandem, *e.* Herringbone.

parlours there are two lines of stalls with the operator(s) working in between. Single-sided tandem parlours always have a milking unit to each stall, but double-tandem parlours usually have a single unit which can serve two stalls situated on opposite sides of the parlour.

Chute parlours are a double tandem type in which the walls of the building form the outsides of the stalls. All the cows on one side have to be changed while the units are in use on the other side, and the only two advantages are cheapness and the fact that the parlour can be contained in a building 10 ft (3 m) wide.

The herringbone parlour is a two-level type in which there are typically 8–20 stalls, with 4–10 units which can work on either side. All the cows on one side are changed while the units on the other side are in operation.

There is less difference between parlours in speed of work than in the ease with which the work is done. Studies of energy used in U.S.A. showed no advantage in having a difference in levels of 12 in–14 in (30–35 cm), but 50 per cent more energy was needed for the abreast type compared with parlours where the difference in levels was 32 in–36 in (81–91 cm). This finding is in accord with practical experience in Britain, and the herringbone parlour should therefore be chosen in preference to the abreast type where practicable. A difference in levels of 30 in (76 cm) is satisfactory for men of average height. The chief merit of the abreast type is that it can often be more easily accommodated in existing buildings.

For herds of 50 or over, there is much to be said in favour of the herringbone. Three units and 6 stalls are adequate for up to 50 cows. For 50 to 100, 4 units and 8 stalls may be used; but where the number is over 70, 5 units and 10 stalls are generally considered more suitable for one-man operation. For large herds, two men are normally needed, and they will require either 8 units and 16 stalls or 10 units and 20 stalls for herds up to 200.

The use of automatic teat-cup removal devices will certainly increase the number of units one man can handle and the number of cows one man can milk.

Rotary parlours have potential advantages for large dairy units which seem likely to influence farmers in their favour as milking work is increasingly automated. The rotary principle makes it possible for some automatic operations to be done at a single point, and this should facilitate economic automation. There are still arguments about such questions as removal of teat-cups at a fixed point at the end of the circuit, and automatic disinfection of teat-cups. Nevertheless it is certain that the routine will be increasingly automated. Automatic cluster removal is already a proven practice and other operations which can be automated with advantage are recording, and transfer of the milk to the dairy. Disinfection of the teats by dipping or spraying is also likely to be increasingly automated.

Washing the udder and taking fore-milk can if desired be carried out in a special stall outside the rotary parlour. Where this is done a high rate of flow of cows into the parlour can be maintained. At the other end of the scale, small machines of the stop–start type cater for 65–70 cows per hour with one-man operation. With these, transfer of the stall has to be positively initiated by the operator. Most of the larger tandems and

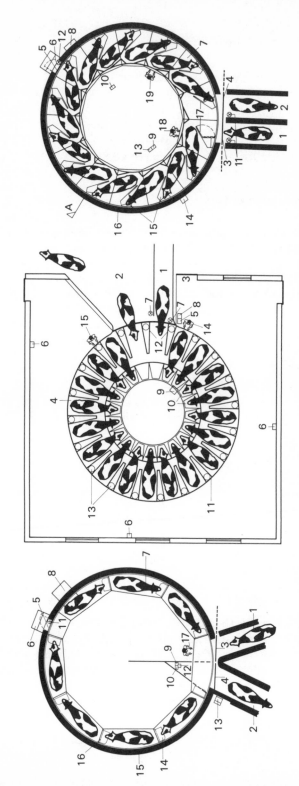

Fig. 13. Types of rotary parlour—*Left:* Tandem. *Centre:* Abreast. *Right:* Herringbone. (Fullwood)

herringbone parlours move continuously, at a speed that can be easily varied. The herringbone and abreast or 'turnstile' types allow more cows to be packed into a given diameter, compared with the tandem arrangement. Thus, for one make, an 18-point tandem needs a 51 ft (15·5 m) building as opposed to a 30 ft (9 m) building for an 18-point herringbone.

Management Factors

In choosing a parlour, account needs to be taken of several management factors, which are outlined below.

1. *Size of herd and calving policy.*
This will determine the maximum number of cows in milk at any time, and thus the number of cows that must be put through the parlour in a given milking time, or number of cows per hour. It may be desirable to consider the possibility of expansion later on.

2. *Average yield of herd.*
The yield largely determines milking time. Average actual milking time is related to average yield approximately as follows:

Yield per Milking		Actual* Milking Time
gallons	*litres*	*min*
$\frac{1}{2}$	2·3	3·3
1	4·5	3·7
$1\frac{1}{2}$	6·8	4·2
2	9·0	4·6
$2\frac{1}{2}$	11·4	5·0
3	13·7	5·5

* Includes allowance of 20% 'over-milking'.

It must be remembered that morning and evening milking are usually unequal, and that with many herds there is a seasonal peak in production. Thus, in a herd that averages 900–1000 gallons (4100–4600 l) several cows may be expected to yield as much as $2\frac{1}{2}$–3 gallons (11–14 l) at a morning milking.

3. *Concentrate feeding policy.*

Where concentrates are fed in the parlour, sufficient time must be allowed for the necessary amounts to be eaten, and the parlour layout and routine must be designed to suit. Average rate of eating cubes is about $\frac{3}{4}$ lb (0·34 kg) per minute. The amount of time a cow has to eat her concentrates is considerably less in a parlour having one stall per unit than it is or can be where there are two stalls per unit. In the former she has only the milking time plus the time taken for the work routine, whereas in a one-man parlour with two stalls per unit, the amount of time available for eating can be almost doubled without unduly reducing the percentage of the operator's time that is spent on the routine work. An advantage of liquid feeding is that it allows a rate of eating of 2–4 lb (0·9–1·8 kg) per minute.

4. *The milking routine time.*

The time needed to perform the necessary milking routine may vary considerably, not only owing to the differences between individual workers, but also according to managerial policy on such points as udder washing and stripping, cleanliness of the cows, etc.

The routine itself differs according to the type of parlour, and particularly according to whether there are one or two stalls per unit.

Parlours With One Stall per Unit

The basic routine in a one stall-per-unit parlour is (1) machine strip, (2) remove and hang cluster, (3) change cows, (4) wash and fore-milk, and (5) put on unit. All these operations are carried out in succession in a single stall, and in a 3-unit 3-stall parlour, similar operations will be carried out in two further stalls before the operator can return to the first cow. Investigations have shown that the total time for this basic routine in farm parlours varies widely. In one study where the total averaged 110 seconds, it was made up as follows: Machine strip, 40 sec; remove and hang cluster, 10 sec; change cows, 20 sec; wash and fore-milk, 30 sec; put on unit, 10 sec.

In the case of 3-unit 3-stall parlours, the minimum actual milking time allowed by this routine is 2×110 sec $= 3$ min 40 sec, and each time a cow is changed there is a period totalling 70 sec during which the machine is not actually milking (Change-over-time).

The number of cows that can be milked per hour can be calculated if the average milking time is known. For example, with a herd that will milk out in 5 minutes per cow, the number of cows that can be milked per hour

$$= \frac{60 \times \text{No. of units}}{\underset{\text{(min)}}{\text{Milking Time}} + \underset{\text{(min)}}{\text{Change-over-Time}}} = \frac{60 \times \text{No. of units}}{5 + 1\frac{1}{6}}$$

With 2 units, the number milked is 19·6 per hour and with 3 units it is 29. Assuming that 3 units are used, the operator is employed for $\dfrac{29 \times 1\frac{5}{6}}{60} \times 100$ per cent $= 90$ per cent of his time, and this would be about what could be expected in practice.

It is considered generally inadvisable to have the operator occupied on milking routine for more than 90 per cent of his time, otherwise he has insufficient time to devote to such matters as keeping check on the condition of individual cows. In the case of the 3-unit parlour considered above, the operator would be occupied 85 per cent of his time if the cows milked out in $5\frac{1}{4}$ minutes, and only 70 per cent of his time if on account of high yields or over-milking, the milking time rose to 7 minutes, when the throughput would fall to 22 cows per hour. With a parlour of this type, the time available for eating concentrates if milking time is $5\frac{1}{4}$ minutes is only about 7 minutes, during which time no more than about 5 lb (2·3 kg) of concentrates can be eaten. Thus, the 3-unit 3-stall parlour may be considered suitable for most herds ranging from 20 to 50 cows, but it is not suitable for high-yielding herds if concentrates need to be fed during milking. When yields are low, there is bound to be some over-milking, but this type of parlour will allow more cows to be milked per hour than the 2-unit 4-stall type.

Fig. 14 shows the relation between number of cows milked per hour and average milking time in parlours having one stall per unit and various numbers of units, assuming routine times

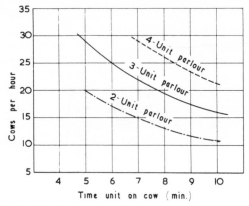

Fig. 14. Relationship between milking time and throughput in one-stall-per-unit parlours with 2, 3 and 4 units, assuming a routine time of 110 seconds.

Source: K. J. Chetwynd.

as detailed above. Where work routine times are reduced to about 0·8 min and a throughput of about 70 cows per hour is achieved, the main saving compared with the times given above for the routine elements is on machine stripping, which is reduced to practically nil. Time spent on the other elements is also reduced by streamlining the work.

Parlours With Two Stalls per Unit

The basic routine for parlours having two stalls per unit is somewhat different. After machine stripping (40 sec), the operator immediately transfers the unit (20 sec) to the opposite cow, which may be called cow A, and which has been waiting already washed for some time. He then returns to the first stall and changes cows (20 sec) and proceeds to wash and fore-milk the new cow which may be called cow B (30 sec), before moving on to the next pair of stalls. The total routine time is the same as for the one-stall-per-unit plant already considered, but the minimum time that must elapse before the operator can return to cow A, is $2 \times 110 + 50$ sec $= 4\frac{1}{2}$ min. Cow B has meanwhile been eating concentrates for this length of time, and still has her own milking time plus the transfer and stripping time in which to finish the ration. With a milking time of 6 minutes she will be able to eat up to 10 lb (4·5 kg) of cubes.

The milking machine is out of use for only 20 sec, so each unit in this type of parlour can milk more cows per hour. In practice, a 2-unit 4-stall parlour is suitable for herds of 30—40 cows where up to $7\frac{1}{2}$ lb (3·4 kg) of concentrates needs to be fed and ample time is allowed for udder washing and stripping. 3-unit 6-stall parlours are suitable for herds of 30—50 cows where yields are high and it is necessary to get the milking done by one man in the shortest possible time.

Standard Times

From the foregoing discussion it is sufficiently evident that a low milking time is only one of the many objectives in choosing and operating a milking parlour. Work study has shown that where it is advantageous, the time for the milking routine can be cut down to well below the 110 sec already considered, by reducing the time spent on stripping or entirely eliminating the job, and by such methods as reducing the time spent on washing udders. By such means, routine times as low as 48 sec can be achieved.

The attainment of low figures depends on many points of detail which must be observed in constructing a parlour. For example, it is essential to have a collecting yard of suitable size, and as explained earlier, some farmers have devised ingenious electric fences which can be moved nearer to the parlour by simply pulling a rope, in order to bring any reluctant cows forward. If the doors of a parlour are easily opened and closed from the pit, e.g. by power operation, the stalls are well designed, the yards are suitable and the cowman always handles the herd quietly, there is no reason why it should often be necessary to leave the pit to fetch animals in. Inside the parlour, ease of feeding and cleaning down are important.

Milking Machine Management

When milking routines have been suitably improved, it is sometimes found that the speed at which cows milk out becomes a limiting factor. As a result of research at the National Institute for Research in Dairying, the following broad conclusions may be drawn.

(1) Allowing the vacuum level to fall below 15 in (38 cm) of mercury will definitely reduce milking rate. Vacuum must not be excessive owing to the likelihood of damage to udders, so should be maintained at 15 in (38 cm) of mercury.

(2) Positive control of rate and ratio of pulsations can improve milking speeds. The best choice is usually 60 pulsations per minute at a wide suction–release ratio, e.g. 3 : 1.

Other important labour-saving developments in milking include automatic cleaning.

Bulk handling of milk from farms to the collecting centre has many advantages. The installation of refrigerated bulk collecting tanks on farms has been actively encouraged by the Milk Marketing Board by means of special loans and also by a premium rate for milk collected in bulk. About 70 per cent of all milk is (1974) handled in this way, but it comes from only about 40 per cent of all producers.

Current rates of premium range from 0·32 pence per gallon (about 0·07 p/l) for bulk tanks of over 350 gal (1400 l) capacity, to 1·20 p/gal (0·27 p/l) for sizes below 150 gal (700 l). These rates apply to an initial 3-year period, after which there is a small continuing premium. The Board also provides a 100 per cent loan for provision of new tanks at an interest rate of 1 per cent above the base bank rate. The loan is repayable over 3 years by deduction from milk cheques. To encourage more of the small producers to switch over to bulk handling, arrangements are in hand to extend the scheme to tank sizes as low as 35 gal (160 l), and where necessary the smaller tanks may be made mobile. One of the objectives of the scheme is to make it unnecessary to collect milk from small-scale producers every day. With refrigerated bulk milk this is unnecessary, and it seems certain that bulk tanks emptied less often than daily will become general practice as soon as collecting depots can undertake the necessary re-equipment.

The capital cost of a bulk tank, including installation and

calibration, ranges from about £1000 for a 150 gallon tank to about £2000 for a 500 gallon one. There is usually an annual service charge after expiry of the guarantee, of about £20 per year. Running cost is in the region of £0·50 per cow per year. The returns for the investment include not only the premiums mentioned but also the very considerable saving of farm labour which continues indefinitely. There can be no doubt that bulk collection will soon become the normal method, since bulk tanks are now within reach of very small producers.

Environment Control in Livestock Buildings

As livestock production is intensified, it becomes increasingly necessary to control the environment in the buildings in which stock are housed. This applies particularly to young stock, but even in buildings for adult cattle there are certain minimum standards which must be maintained if the stock are to thrive. The main functions of any environment control system are to maintain temperature within prescribed limits and to effect air changes, in order to maintain a supply of fresh air, and to remove the large amounts of water vapour produced by the stock. Ventilation needs to be regulated in such a way that draughts are avoided, and there are no parts of the building where condensation of moisture occurs in cold weather. One of the big economic problems of the future is how to secure technical efficiency in environment control without excessive capital cost. One of the first considerations is air temperature. In Britain, provision of optimum temperatures usually consists of either warming the air to increase temperature, or providing ventilation to reduce it. As a general rule, it is only necessary to provide artificial heat for young stock and adult poultry. Other adult stock can normally provide sufficient heat to keep themselves warm in a well-designed building. Thermal insulation of the building may be an important design feature where positive control of the environment is attempted.

The general problem of how to find the most economic system of environment control for particular types of stock is a complex one. In the first place, it is necessary to establish what are optimum conditions, and what decrease in production is likely to result from specified departures from this optimum.

Information on this can only be accumulated gradually, because of the complex nature of animal health and disease. For example, a particular type of adverse environment such as a short period of cold, combined with inadequate ventilation, may have little effect on healthy cattle, but may be quickly fatal to others already suffering from a disease affecting respiration.

At present, decisions have to be made without adequate scientific and economic information on the various possibilities. This is partly because the provision of control equipment has developed much faster than controlled experimentation to test methods of use. All that a farmer can do at present is to calculate what provision of various kinds of control equipment will cost annually, and relate this to the margin between production costs and value of the product. For example, if the installation of environment control equipment for a particular enterprise will cost £1000 and fuel and electricity costs for operating it will be £100 per year, total annual cost, allowing for a 10-year life, will be approximately £150 for fixed costs plus £100 for running costs, i.e. £250 per year. This could be a small price to pay to ensure a profitable enterprise, but would be a serious waste of money if it had little or no effect on production.

Generally, it pays to provide full control of environment for all kinds of young stock kept intensively, and partial control for older calves and poultry. For the latter, and for housing adult cattle, pigs and sheep, it is necessary to compare systems as a whole. A particular design of house with controlled (fan) ventilation often has to be compared with a completely different method of housing based on natural ventilation. For adult beef cattle in British conditions it would be difficult in present circumstances to find any economic justification for providing anything more than simple housing, sufficient to avoid wetting of the skin and draughts.

'Automatic' control of environment covers a very wide range of procedures. At the present stage of development, when control equipment is being applied in many new ways, the most practical approach may be to specify such measurable factors as temperature, number of air changes and operating cost, and choose a supplier who will undertake to meet the specification cheaply.

Automation in Livestock Tending

The scope for economic application of automation in livestock tending is vast. Controls which can at present be economically automatic in favourable circumstances include control of environment in buildings for poultry, pigs and young calves; feeding of poultry and pigs; feeding of concentrates to dairy cows in proportion to the amount of milk given, and cleaning of pipe-line milking equipment.

The economy of providing automatic control of environment has been greatly improved by the introduction of a simple speed control on electric motors from full speed down to about one-tenth of full speed. This makes it very much cheaper to achieve control in buildings such as calf houses and all kinds of poultry buildings.

Automation in stock feeding has already been briefly discussed. It not only saves labour, but generally does the job better than the best of men can in practice achieve when handling a large number of animals. When using tower silo unloaders, variations in compactness of the silage in different parts of the silo tend to result in a very poor performance by the unloader. Apart from the occasional necessity to climb the tower in order to clear a blockage, the necessity to stand by the winch in order to make frequent adjustments leads to low efficiency in the use of labour. Automatic control of the unloader can help to overcome such difficulties. Automatic metering and automatic recording of the quantities of grain and fodder fed to livestock will become increasingly necessary. It is not usually sensible to try to evaluate the provision of automatic control equipment simply in terms of labour saving. Often it is necessary to take account of the fact that automation is essential to a new production technique, and that the enterprise itself would not be practicable without it. The costs of automation are easily calculable on the usual basis, but returns are closely linked with success of the particular production enterprise.

Typical capital costs of automatic controls for stationary use are reasonable in relation to what they achieve. In future it can be expected that a very large part of the stockman's work will consist of pressing buttons to start an operation, and then leaving the machine to get on with the job unattended.

List of References

1. *The Changing Structure of Agriculture.* H.M.S.O., London (1970).
2. *Annual Review of Agriculture, 1973.* Cmnd. 5254. H.M.S.O., London (1973).
3. *Some Aspects of the Economic Utilization of Farm Machinery.* B. M. Camm. *J. Proc. Instn. Agric. Engnrs.* 20. 4 (1964) 151.
4. *Economic Aspects of Fuel and Power in Agriculture.* F. G. Sturrock. British Association for Advancement of Science. Cambridge. September 1965.
5. *Machinery Sharing in England and Wales.* Farm Mechanization Studies No. 19. M.A.F.F., A.D.A.S., London (1972).
6. *The Farm as a Business. Introduction to Management.* H.M.S.O., London.
7. *Variation in the Repair Costs of Tractors, Combine Harvesters and Balers.* A. H. Gill. Misc. Study No. 50. Reading University, Dept. of Agricultural Economics and Management (1971).
8. *At the Farmer's Service.* Ministry of Agriculture. Available free at the Ministry's Local Offices.
9. *Farm Management Handbook 1973.* Universities of Bristol and Exeter. Agricultural Economics Research Unit. Bristol (1973).

10. *Farm Management Pocketbook.* 5th Edition. J. S. Nix. Wye College, Ashford, Kent (1972).
11. *Farm Planning Data, 1966.* J. B. Hardaker. Cambridge University Farm Economics Branch (1966).
12. *Work Study.* British Institute of Management. London (1964).
13. *Farm Machinery.* 8th Edition. C. Culpin. Crosby Lockwood Staples. London (1969).
14. *Wheeled and Tracklaying Tractors.* A.D.A.S., N.I.A.E. Study of Utilization, Performance, and Tyre and Track Costs, 1969–70. Farm Mechanization Studies No. 20. M.A.F.F., A.D.A.S., London (1972).
15. *Irrigation.* Ministry of Agriculture Bulletin No. 138. H.M.S.O., London.
16. *Cereals Without Ploughing.* A.D.A.S. Profitable Farm Enterprises Booklet 6. M.A.F.F. (1974).
17. *Farm Grain Drying and Storage.* Ministry of Agriculture Bulletin No. 149. H.M.S.O., London.
18. *The Utilization and Performance of Combine Harvesters, 1969.* Farm Mechanization Studies No. 18. M.A.F.F., A.D.A.S. (1970).
19. *The Utilization and Performance of Potato Harvesters, 1971.* Farm Mechanization Studies No. 24. M.A.F.F., A.D.A.S. (1972).
20. *Bulk Storage of Potatoes in Buildings.* Ministry of Agriculture Bulletin No. 173. H.M.S.O., London.
21. *The Utilization and Performance of Sugar Beet Harvesters, 1973.* Farm Mechanization Studies No. 26. M.A.F.F., A.D.A.S. (1974).
22. *Farm Buildings.* Vol. 1. J. B. Weller. Crosby Lockwood Staples. London (1965).
23. *Feeding of Cattle.* Farm Electrification Handbook No. 19. Electricity Council. London.
24. *Feeding of Pigs and Poultry.* Farm Electrification Handbook No. 20. Electricity Council. London.
25. *Machine Milking.* Ministry of Agriculture Bulletin No. 177. H.M.S.O., London.
26. *Principles of Farm Machinery.* 2nd Edition. R. A. Kepner, Roy Bainer and E. L. Barger. AVI Publishing Co. Westport, Conn. (1972).

Appendix

Table A.1. Effective Capacities of Field Equipment
(i) Soil-working Implements

Implement	Normal working speed m.p.h.	Typical Implement working width ft	depth in	draft lb	field effici-ency %	Working Rates acres per hour spot*	overall	Acres per day (‖) (once over)
Plough 2F G.P.†	3½	2	6	1100	80	0·8	0·6	4
Plough 3F G.P.†	3½	3	6	1650	80	1·3	1·0	7
Plough 4F G.P.‡	3½	4	6	2200	80	1·7	1·4	10
Plough 6F G.P.§	3½	6	6	3300	80	2·6	2·0	15
Plough 1F Deep†	3	1¼	10	1500	80	0·5	0·4	3
Plough 2F Deep ‡	3	2½	10	3000	80	0·9	0·7	5
Plough 3F Deep ‡	3	3¾	10	4500	80	1·4	1·1	8
Plough 4F Deep ¶	3	5	10	6000	80	1·8	1·4	10
Plough 5F Semi-digger ¶	4	5¾	10	7000	80	2·8	2·2	16
Rotary cultivator †	2	5	4	—	85	1·2	1·0	7
Rotary cultivator †	1½	5	6	—	85	0·9	0·7	5
Rotary cultivator ‡	2½	5	4	—	85	1·5	1·3	9
Rotary cultivator ‡	2	5	6	—	85	1·2	1·0	7
Tine cultivator Heavy ‡	3½	7	6	3000	85	3·0	2·5	18
Tine cultivator Heavy §	3½	10	6	4500	85	4·4	3·7	26
Tine cultivator Heavy ¶	3½	12	8	7000	85	5·2	4·5	32
Spring-tine cult-harrow †	5½	9	3	900	85	5·5	4·6	32
Spring-tine cult-harrow ‡	5	13	3	1300	85	8·0	6·8	48
Spring-tine cult-harrow §	5	20	3	2000	85	12·0	10·0	70
Harrow Light †	4	10	2	500	85	4·2	3·5	25
Harrow Disc †	3½	7	3	750	85	3·0	2·5	18
Harrow Disc ‡	3½	8	4	1000	85	3·4	2·8	20
Harrow Disc §	4	10	4	1300	85	4·3	3·6	25
Harrow Disc Heavy ¶	4	10½	6	4500	85	5·2	4·5	32
Roll †	4	16	—	600	85	7·8	6·5	45
Tractor hoe †	2	8	—	—	80	1·9	1·5	10
Down-row thinner †	2	8	—	—	80	1·9	1·5	10
3-row ridger or scuffler †	3	7	—	—	80	2·5	2·0	14

* Spot working rate is the number of acres that would be covered by the implement travelling in a straight line at a steady speed. i.e. it is the product of the forward speed and the working width.
    † Small-medium (31–45 h.p.) tractor.
    ‡ Medium (46–60 h.p.) tractor.
    § Large-medium (61–80 h.p.) tractor.
    ¶ Large (over 80 h.p.) tractor.
    ‖ It is assumed that day consists of 8 working hours, but that only 7 hours on average are spent on the actual job, owing to travelling time, etc.

Table A.1. Effective Capacities of Field Equipment
(ii) Distribution and Drilling

Equipment	Hopper size cwt	Appli-cation rate cwt/acre	Gang size	Typical Implement working width ft	field effici-ency* %	Normal working speed m.p.h.	Working Rates acres per hour spot	overall*	Acres per day
Spinner	6	3	1	20	50	5	12	6·0	42
Spinner	6	10	2	20	40	5	12	4·8	34
Spinner bulk handling	30	3	1	20	75	5	12	9·0	63
Spinner bulk handling	30	10	1	20	50	5	12	6·0	42
Full width distributor	6	3	1	8	65	4	3·8	2·5	18
Full width distributor	6	10	2	8	55	4	3·8	2·0	14
Full width distributor	12	3	2	17	65	4	8·1	5·2	36
Full width distributor	12	10	2	17	45	4	8·1	3·6	25
Combine drill	4 seed	1¼	2	8	60	5	4·8	2·9	20
	5 fert	3	—	—	—	—	—	—	—
Corn drill	4	1¼	1	8	70	5	4·8	3·4	24
Corn drill	15	1¼	1	13	70	5	8·0	5·6	40
Spacing drill 5-row	—	—	1	8	60	2	1·9	1·1	8
Farmyard manure spreader, wheel drive	40	200	1	7	†	3	2·5	†	†
Farmyard manure spreader, p.t.o.	80	200	1	8	†	5	5·0	†	†
	Tank cap. gal	Gal per acre							
Slurry tanker	300	3000	1	—	—	—	—	0·3	2
Slurry tanker	700	3000	1	—	—	—	—	0·7	5
Field crop sprayer	80	20	1	15	50	4	7·2	3·6	25
Field crop sprayer	100	20	1	40	40	4	19·4	7·7	55
Row crop spray and drill, 5-row	—	—	1	8	40	2	1·9	0·8	6

* The field efficiency and overall rates allow for carting from store in good working conditions at a short (¼ mile) transport distance.
† Depends on organization.

Table A.1. Effective Capacities of Field Equipment
(iii) Corn, Hay and Silage Harvesting

Machine	Working width ft	Normal working speed m.p.h.	Field efficiency %	Working Rates acres per hour spot	overall	Gang size	Acres per day
Combine harvester * may be p.t.o.	6	$1\frac{1}{2}$–3	75	1–2·1	$\frac{3}{4}$–$1\frac{1}{2}$	2	8†
Combine harvester * medium capacity	10	2–4	75	2·4–4·8	$1\frac{3}{4}$–$3\frac{1}{2}$	2	16†
Combine harvester * high capacity	12	2–4	75	2·9–5·7	2–4	2	20†
Combine harvester * high capacity	14	2–4	75	3·3–6·7	$2\frac{1}{2}$–5	2–3	25†
Combine harvester,* giant	20	2–4	75	4·8–9·6	$3\frac{1}{2}$–7	3	35†
Pick-up baler, straw *	10	4	60	4·8	2·9	1	20†
Pick-up baler, hay	10	4	50	4·8	2·4	1	15
Mower, finger bar	5	$3\frac{1}{2}$	75	2·1	1·5	1	10
Mower, flail	5	4	85	2·4	2·0	1	15
Mower, rotary	5	6	80	3·6	2·9	1	20
Tedder, 1-row	5	5	85	3·0	2·5	1	15 §
Tedder, 2-row	10	5	85	6·0	5·0	1	30 §
Tedder, 3-row	15	5	85	9·0	7·5	1	45 §
Swath turner/siderake	10	5	85	6·0	5·0	1	30 §
Forage harvester, flail	4	3	65 ‡	1·4	0·9	2	6
F.H. 'Full-chop (pick-up)'	10	$2\frac{1}{2}$	65 ‡	3·0	2·0	2	12

* Assume 20-day harvest for normal cropping.
† Reduce by 25 per cent in North and West.   ‡ Depends on team.
§ Effective day length for operation reckoned at 6 hours.

Table A.1. Effective Capacities of Field Equipment
(iv) Specialist Potato and Sugar Beet Machinery

Machine	Working width ft	Normal working speed m.p.h.	Field efficiency %	Working rates acres per hour spot	overall	Gang size	Acres per day
Potato Planters							
Hand-fed 2-row	5	$1\frac{1}{4}$	60	0·8	0·5	4	$3\frac{1}{2}$*
Hand-fed 3-row	$7\frac{1}{2}$	$1\frac{1}{4}$	60	1·2	0·7	5	5 *
Hand-fed 4-row	10	$1\frac{1}{4}$	60	1·6	1·0	6	7 *
Automatic 2-row							
with fertilizer	5	3	60	1·8	1·0	2	7 †
oscillating feed	6	5	60	3·6	2·2	2	15
Potato Elevator							
Digger							
1-row	$2\frac{1}{2}$	2	70	0·6	0·4	6 + 12	$2\frac{1}{2}$ ‡
2-row	5	2	80	1·2	0·9	1 **	** §
Potato Harvester							
(main crop) 1-row	$2\frac{1}{2}$	$1\frac{1}{4}$	70	0·4	0·28	6–8	2 ¶
(unmanned) 2-row	6	$1\frac{1}{2}$	70	1·0	0·7	4·6	5
Sugar-Beet Harvester							
1-row side elevator	$1\frac{2}{3}$	3	75	0·6	0·45	3	$3\frac{1}{4}$ ‖
1-row tanker	$1\frac{2}{3}$	3	75	0·6	0·45	2	$3\frac{1}{4}$ ‖
2-stage, 3-row	5	3	70	1·8	1·25	4	$8\frac{1}{2}$ ‖
3-stage, 5-row	$8\frac{1}{3}$	$2\frac{1}{2}$	70	2·5	1·75	6	12 ‖

* Rate of work reduced by 33% when planting chitted seed.
† Some makes not suitable chitted seed.   ‡ Working $6\frac{1}{2}$ hours a day.
§ Sufficient for day lifted in 2–3 hours.   ¶ Partial mechanical separation.
‖ Less late in season.   ** Digger works independently.

Table A.1. Effective Capacities of Field Equipment
(v) Transplanters

	Typical Spacing		Typical Performance, 1 man/unit		
Crop	In row in	Between rows in	Forward speed m.p.h.	Plants per hour per unit	Acres per hour (overall for) 2-row planter
Leeks	4	20	0·2	2500	0·15
Celery	7½	60	0·3	2700*	0·30
Cabbage	14	21	0·45	—	0·15
Brussels Sprouts	32	36	0·9	—	0·50

* One extra man per 2 rows.

Table A.1(vi). Calculation of Working Rate in the Field

Working rate depends on forward speed, working width and field efficiency. Typical field efficiencies are:

Tillage implements, 75—87 per cent.
Distribution and drilling, 50—75 per cent.
Corn, hay and silage harvesting, 60—85 per cent.
Low volume field crop spraying, 40 per cent.
Potato and sugar-beet machinery, 60—80 per cent.

Some details are given in tables A.1(i) to A.1(iv).

Working rate = field efficiency × area that would be covered if the equipment could work at a constant forward speed, with no stops or turns. The figures below give approximate theoretical acreages covered per hour at constant forward speed and working width, and 100% efficiency.

Working width in	Speed m.p.h.							
	1	2	3	4	5	6	7	8
10	0·1	0·2	0·3	0·4	0·5	0·6	0·7	0·8
20	0·2	0·4	0·6	0·8	1·0	1·2	1·4	1·6
30	0·3	0·6	0·9	1·2	1·5	1·8	2·1	2·4
40	0·4	0·8	1·2	1·6	2·0	2·4	2·8	3·2
50	0·5	1·0	1·5	2·0	2·5	3·0	3·5	4·0
60	0·6	1·2	1·8	2·4	3·0	3·6	4·2	4·8
70	0·7	1·4	2·1	2·8	3·5	4·2	4·9	5·6
80	0·8	1·6	2·4	3·2	4·0	4·8	5·6	6·4
90	0·9	1·8	2·7	3·6	4·5	5·4	6·3	7·2

Table A.1(vi)—*continued*

Working width in	1	2	3	Speed m.p.h. 4	5	6	7	8
100	1·0	2·0	3·0	4·0	5·0	6·0	7·0	8·0
120	1·2	2·4	3·6	4·8	6·0	7·2	8·4	9·6
140	1·4	2·8	4·2	5·6	7·0	8·4	9·8	11·2
160	1·6	3·2	4·8	6·4	8·0	9·6	11·2	12·8
180	1·8	3·6	5·4	7·2	9·0	10·8	12·6	14·4

To measure forward speed, stick two pegs in the ground 88 ft apart, and time the outfit over this distance.

$$\text{Speed, m.p.h.} = \frac{60}{\text{time in seconds to travel 88 ft}}$$

Rough estimate method, applicable to low speeds only. Walk alongside the equipment, counting the number of paces of 1 yard each taken in 20 seconds.

$$\text{Speed, m.p.h.} = \frac{\text{Number of paces taken}}{10}$$

Example. A 3-furrow tractor plough with 10-inch furrows is travelling at 5 m.p.h.

Theoretical working rate from Table A.1(vi) is 1·5 acres/hr. It is working in a small, irregular field; so field efficiency is assessed at 75 per cent.

Estimated working rate $= \dfrac{75}{100} \times 1\cdot5 = \underline{\underline{1\cdot1}}$ acres per hour.

Daily rate for an 8-hour day is likely to be less than 8 × 1·1 acres, due to time taken in travelling to and from work, etc.

Table A.2(i). Estimated* Useful Life (Years) of Power-Operated Machinery in Relation to Annual Use

Equipment	Annual Use (hours)				
	25	50	100	200	300
Group 1: Ploughs, Cultivators, Toothed harrows, Hoes, Rolls, Ridgers, Simple potato planting attachments, Grain cleaners.	12+	12+	12+	12	10
Group 2: Disc harrows, Corn drills, Binders, Grain drying machines, Food grinders and mixers.	12+	12+	12	10	8
Group 3: Combine harvesters, Pick-up balers, Rotary cultivators, Hydraulic loaders.	12+	12+	12	9	7
Group 4: Mowers, Forage harvesters, Swath turners, Side-delivery rakes, Tedders, Hedge cutting machines, Semi-automatic potato planters and transplanters, Unit root drills, Mechanical root thinners.	12+	12	11	8	6
Group 5: Fertilizer distributors, Combine drills, Farmyard manure spreaders, Elevator potato diggers, Spraying machines, Pea cutter-windrowers.	10	10	9	8	7
Miscellaneous: Beet harvesters	11	10	9	6	5
Potato harvesters	—	8	7	5	—
Milking machinery	—	—	—	12	10

	Annual Use (hours)					
	500	750	1000	1500	2000	2500
Tractors	12+	12	10	7	6	5
Electric motors	12+	12+	12+	12+	12	12

* The above figures are the author's estimates and are suggested as a general guide only. See pp. 28–33.

Table A.2(ii). Depreciation: Average Annual Fall in Value. Per Cent of New Price

Frequency of renewal Years	Complex. High Depreciation Rate. e.g. potato harvesters, mobile pea viners, etc.	Established machines with many moving parts, e.g. tractors, combines, balers, forage harvesters	Simple equipment with few moving parts, e.g. ploughs, trailers
	%	%	%
1	34	26	19
2	$24\frac{1}{2}$	$19\frac{1}{2}$	$14\frac{1}{2}$
3	20*	$16\frac{1}{2}$*	$12\frac{1}{2}$
4	$17\frac{1}{2}$†	$14\frac{1}{2}$	$11\frac{1}{2}$
5	15‡	13†	$10\frac{1}{2}$*
6	$13\frac{1}{2}$	12	$9\frac{1}{2}$
7	12	11	9
8	11	10‡	$8\frac{1}{2}$†
9	(10)	$9\frac{1}{2}$	8
10	$(9\frac{1}{2})$	$8\frac{1}{2}$	$7\frac{1}{2}$‡

* Typical frequency of renewal with heavy use.
† Typical frequency of renewal with average use.
‡ Typical frequency of renewal with light use.

Source: V. Baker, Bristol University.

Table A.3. Estimated Annual Cost of Spares and Repairs as a Percentage of Purchase Price* at Various Levels of Use

	Approximate Annual Use (hours)				Additional use per 100 hours ADD
	500	750	1000	1500	
	%	%	%	%	%
Tractors	5	6·7	8·0	10·5	0·5

	Approximate Annual Use (hours)				Additional use per 100 hours ADD
	50	100	150	200	
	%	%	%	%	%
Harvesting Machinery					
Combine Harvesters, self-propelled and engine driven	1·5	2·5	3·5	4·5	2·0
Combine Harvesters, p.t.o. driven, metered-chop forage harvesters, pick-up balers, potato harvesters, sugar-beet harvesters	3·0	5·0	6·0	7·0	2·0
Other Implements and Machines					
Group 1: Ploughs, Cultivators, Toothed harrows, Hoes, Elevator potato diggers } Normal soils	4·5	8·0	11·0	14·0	6·0
Group 2: Rotary cultivators, Mowers, Binders, Pea-cutter-windrowers	4·0	7·0	9·5	12·0	5·0
Group 3: Disc harrows, Fertilizer distributors, Farmyard manure spreaders. Combine drills, Potato planters with fertilizer attachment, Sprayers, Hedge-cutting machines	3·0	5·5	7·5	9·5	4·0
Group 4: Swath turners, Tedders, Side-delivery rakes, Unit drills, Flail forage harvesters, Semi-automatic potato planters and transplanters, Down-the-row thinners	2·5	4·5	6·5	8·5	4·0
Group 5: Corn drills, Milking machines, Hydraulic loaders, Simple potato planting attachments	2·0	4·0	5·5	7·0	3·0
Group 6: Grain driers, Grain cleaners, Rolls, Hammer mills, Feed mixers, Threshers	1·5	2·0	2·5	3·0	0·5

* When it is known that a high purchase price is due to high quality and durability, or a low price corresponds to a high rate of wear and tear, adjustments to the figures should be made. See p. 37.

Table A.4. Annual Labour Requirements for Different Enterprises (Standard Man Days)

Crops and grass	Standard man days per acre after J. S. Nix[10]	M.A.F.F. basis for official statistics
Wheat and barley	1·5*	2
Oats	1·5*	3
Potatoes	12	15
Sugar beet	5	10
Vining peas	2	3
Threshed beans	2	3
Threshed peas	2	3
Herbage seeds	1	—
Hops	50	70
Feed roots, cut	8	9–11
Feed roots, folded	3	—
Kale, cut	6	—
Kale grazed	1	1·5
Hay/silage 1 cut	1·5	0·5–1·25
Hay/silage 2 cuts	2	—
Grazing only	0·5	0·5
Bare fallow	1	0·5

Livestock	Standard man days per head	
Dairy cows: parlour	7	10
Dairy cows: cowshed	10	10
Bulls (for service)	4	6
Beef cows	2·5	3
Barley beef	3	—
15–18 month beef	3	—
Other cattle over 2 years	2·5	2·5
Other cattle 1–2 years	2	2·5
Other cattle ½–1 year	1†	2·5
Calves ½–1 year (not suckled)	2†	2·5
Ewes and rams	0·5	0·7
Other sheep over 6 months	0·3	0·2
Sows	4	4
Boars	2	4
Other pigs over 2 months	0·75	1
Laying birds, intensive	0·1	0·1
Broilers produced	0·01†	0·05
Turkeys	0·1	0·1

* Including straw. If straw burnt or ploughed in, 1 man day.
† SMDs based on numbers produced annually.
When assessing regular staff required:
Either assume 300 SMDs supplied by stockmen and 250 by other workers, *or* add 15 per cent for general maintenance and assume 300 SMDs are provided by all workers.

Table A.5. Size of Livestock Enterprises Per Man Unit

| | | Full-time unit. Animals/man | | |
| | | | Cows and followers | |
Enterprise	System	Cows only	1 : 1 ratio	Notes
Dairy cows	Cowshed	30–42	23–33	Bucket plant
	Yard and parlour		60 cows	Abreast, tandem or chute
	Yard and herringbone or rotary parlour	100		Efficient layout

		Full-time unit. Animals/man	
Beef cattle	Rear single suckled	80–100 cows	
	Rear double suckled	50–60 cows	
	Rear multiple suckled	20–30 cows	7–10 calves per lactation
	Fatten strong stores in yards	120–150 animals	Traditional
	Yarded stores	300 animals	Well designed yards
	Intensive beef	600 animals	Including rearing
Sheep	Arable flock	400–500 ewes	Plus help for lambing and dipping
	Grassland flock	600 ewes	
Pigs	Weaner production	120 sows	Efficient layout
	Breed and fatten to bacon weight, hand feeding	62 sows	Efficient layout
	Mechanical feeding	72 sows	Efficient layout
	Fattening pigs, hand feeding	850 pigs	
	Fattening pigs, mechanical feeding	1500 pigs	Assistance for slurry carting
Poultry – specialized units	Layers, battery	8 000	Non-mechanical system
	Table birds	30 000–50 000 birds per year	Mechanical feeding

Sources: Nottingham University and Cambridge University Departments of Agricultural Economics.

Table A.6(i). Seasonal Labour Requirements. Man Hours Per Acre and Per Hectare of Crop. Premium Applicable to Most Farms with More than 300 Acres (122 ha) of Arable Land

Crop	Winter wheat			Spring barley		
	Man hours			Man hours		
Month	per acre	per ha	Operations	per acre	per ha	Operations
Oct.	2·0	5·0	Plough (⅔), cultivate drill, harrow	0·3	0·7	Plough
Nov.	0·4	1·0	Plough (10–15%) cult drill, harrow	0·6	1·5	Plough
Dec.	—	—	—	0·3	0·7	Plough
Jan.	—	—	—	—	—	
Feb.	—	—	—	—	—	See March
Mar.	0·2	0·5	Top dress, roll	1·7	4·3	Cultivate, drill. Nearly half in Feb. on light land
Apr.	0·4	1·0	Top dress, roll, spray	0·2	0·5	Roll
May	—	—	—	0·2	0·5	Spray
June	—	—	—	—	—	—
July	—	—	—	—	—	—
Aug.	0·9	2·2	Grain harvest	1·2	3·0	Grain harvest
	1·2	3·0	Straw harvest	1·3	3·2	Straw harvest
Sept.	0·3	0·7	Grain harvest	0·9	2·3	Straw harvest
	1·0	2·5	Straw harvest	—	—	—
	0·6	1·5	Plough (20–25%) cult drill, harrow	—	—	—
Total excl. barn work	7·0	17·4		6·7	16·7	

Winter barley
As winter wheat except ploughing unlikely to start before cereal harvest. Harvest end July–beginning August.

Winter oats
As winter wheat except drilling complete earlier (mid-September to mid-October). Harvest earlier. (First half of August.)

Spring wheat
As spring barley except drilling 2 weeks earlier (completed in March). But variety Opal an exception (April drilling suitable). Harvest August, 2 weeks later than barley.

Spring oats
As spring barley except drilling 2 weeks earlier. Harvest later than spring barley but earlier than spring wheat.

Crop	Maincrop potatoes			Sugar beet		
	Man hours			Man hours		
Month	per acre	per ha	Operations	per acre	per ha	Operations
Oct.	16·2	40·5	80% of harvest, ½ burn off	4·0	10·0	45% harvest + loading
Nov.	2·9	7·3	Clamp work, ¾ plough	4·9	12·3	45% harvest + loading, ¾ plough
Dec.	0·3	0·7	¼ plough	0·6	1·5	¼ plough + loading
Jan.	—	—	—	0·2	0·5	Loading
Feb.	—	—	—	—	—	—
Mar.	1·4	3·5	All fertilizer, ½ cultivation, ¼ plant	2·3	5·7	Fertilizer. Most of cultivation, some drilling
Apr.	2·0	5·0	½ cultivation, ¾ plant	2·0	5·0	Some cultivation. Most of drilling
May	0·4	1·0	¼ after-cultivation. Spray	1·1	2·7	½ thinning. 40–45% tractor hoeing

Table A.6(i)—*continued*

| | Maincrop potatoes | | | Sugar beet | | |
| | Man hours | | | Man hours | | |
Month	per acre	per ha	Operations	per acre	per ha	Operations
June	—	—	$\frac{1}{2}$ after-cultivation	1·1	2·7	$\frac{1}{2}$ thinning. 40–45% tractor hoeing
July	0·5	1·2	$\frac{1}{4}$ after-cultivation, 2 blight sprays	0·5	1·3	Rest of tractor hoeing. Spraying (aphis)
Aug.	0·3	0·7	1 blight spray	—	—	—
Sept.	4·1	10·3	20% harvest, $\frac{1}{2}$ burn off	1·0	2·5	10% harvest + loading
Total	28·1	70·2		17·7	44·2	

Notes. 2-row automatic planter. Herbicide sprays. Minimal after-cultivation. Mechanical harvester. Figures exclude casual labour which may be 15 hours/acre (37 h/ha) for work on harvester.

Notes. Two herbicide sprays, drill to stand, no hand hoeing. 3 tractor hoeings. Harvest by 1-man tanker or 2-row. Hydraulic loader used. 20% less harvest time on light land.

| | Grass conservation – hay 1·6 ton/acre (4 ton/ha) | | | Grass conservation – silage 6 ton/acre (15 ton/ha) | | |
| | Man hours | | | Man hours | | |
	per acre	per ha	Operations	per acre	per ha	Operations
June $\frac{2}{3}$	0·7	1·8	Mow	May $\frac{1}{3}$	0·7	1·8	Mow
July $\frac{1}{3}$	0·7	1·8	Ted and turn	June $\frac{2}{3}$	4·1	10·2	Load, cart, clamp
	0·5	1·2	Bale				
	2·9	7·2	Cart				
Total	4·8	12·0		Total	4·8	12·0	

Source: J. S. Nix, *Farm Management Pocketbook*, 5th Edition, 1972, which see for further crops and details.

Table A.6(ii). Seasonal Distribution of Labour Requirements (expressed as monthly percentages of total)

Enterprise	Seasonal distribution of labour, per cent											
	Jan.	Feb.	Mar.	Apr.	May	June	July	Aug.	Sept.	Oct.	Nov.	Dec.
Cereal, winter	—	8	8	8	—	—	—	17	17	17	17	8
Cereal, spring	—	—	20	20	—	—	—	20	20	10	10	—
Potatoes, main crop	3	4	6	12	3	3	1	—	12	41	12	3
Sugar beet	—	—	1	7	27	13	5	—	7	27	13	—
Kale, folded	—	—	12	44	32	12	—	—	—	—	—	—
Hay	—	—	10	10	—	40	30	10	—	—	—	—
Silage	—	—	10	10	40	30	10	10	—	—	—	—
Grazing	—	—	17	17	—	16	17	17	16	—	—	—
Dairy cows (shed)	10	10	10	10	6	6	6	6	6	9	10	11
Dairy cows (parlour)	10	10	9	8	7	7	7	7	8	8	9	10
Other cattle Over 1 year	15	15	14	5	3	3	3	3	5	10	10	14
Other cattle Under 1 year	11	9	9	6	6	3	3	3	11	14	14	11
Sheep	5	5	18	11	11	16	6	5	6	5	6	6
Pigs and poultry	Fairly even distribution depending on system											

Source: Farm Business Data, University of Reading, Dept. of Agric. Economics and Management, 1972.

Table A.7 Estimate of Days Available for Field Work in the Eastern Counties [11]

Month	Days available	Season	Days available
Mar.	20	Early spring	40
Apr.	22		
May	24	Late spring	30
June	25		
		Early summer	27
July	24		
		Mid summer	30
Aug.	24		
		Late summer	28
Sept.	22		
Oct.	19	Early autumn	31
Nov.	16		
		Late autumn	30
Dec.	14		
Jan.	14		
		Winter	21
Feb.	13		

Source: Farm Planning Data 1966. J. B. Hardaker, Camb. Univ. F.E.B., October 1966.

Table A.8. Labour Requirements for Some Gang-Work Operations *

Operation	No. of Workers Reg.	No. of Workers Casual	Equipment	Method	Acres per 8 hour day	Notes	Piece-work or contract Cost
Hand pick early potatoes. 5-ton crop	1[1]	8 W	Elevator digger	Hand riddle in field	1·0	(1) 2 Extra men loading 1–2 hours. Women working 6 hours	
Harvest early potatoes. 10-ton crop	7		Complete harvester		1·45	Including loading, July–August	
Hand-pick main crop potatoes	4	6[2]	Elevator digger, 2 trailers	Breadth system	1·5	(2) 6 Experienced pickers working 6 hours	Picking and loading. £30–£40 per acre
	6	12	Elevator digger, 2–3 trailers	Stint system	1·65	All working 6 hours	
	3	10	Elevator digger, boxes, loader, 1 trailer	Box-handling system	1·2	Pickers working 6 hours. Regulars 8 hours	
	5	18	Elevator digger, boxes, loader, 2–3 trailers		2·2		
Harvest main-crop potatoes	8		Complete harvester, 2 trailers	Side-loading into trailer	1·8	Working 8 hours per day	
Riddling potatoes	3–6		Power riddle		—	3 Tons per man per day for good sample. Output drastically reduced if poor sample	
Chopping out sugar beet	1		Long hoe		0·8–1·0	May–June 24 days available	
Singling sugar beet		1		By hand	0·6–1·0		
Chopping out and singling sugar beet	1				0·4		£15–£20 per acre

Operation	Men		Machine	Transport	Rate	Output/conditions	Notes
Sugar beet harvesting	3		Side-elevator harvester, 2 trailers	Cart to heap	2·0–2·25	September–November. Good conditions	Hire of harvester and driver, from £25 per acre
Sugar beet loading	1		Front loader		—	½ hour per 8-ton load	
Cutting vining peas	1		6 ft Pea cutter		3–5·3	Output varies with conditions up to about 20 acres per day (limited by viner)	
Loading vining peas	3		Green-crop loader	To lorries for factory vining	12	Early: 6 tons/acre	Depends mainly on weight of vine
					8	Mid-season: 10 tons/acre	
					5·7	Late: 14 tons/acre	
Vining	8–11		Static viner		3·2–4·0	Early	Includes cutting and carting but not transport of peas to factory
					2·3–3·2	Mid-season	
					1·8–2·3	Late	
Vining	4		Mobile viner		2·4–4·0	Viners usually worked 24 hour/day by two teams of 4 men each, giving total of up to 10 acres per day. Cutting labour included, not transport to factory	
Harvesting early carrots	1	12	Potato elevator digger	Lift as early potatoes	1	5-ton crop	
Machine harvesting main crop carrots	3		Complete harvester, 2 trailers	Side-loading	2	15-ton crop	
Brussels sprouts picking	1		—	Hand picking	·05	Picked over 3–5 times per season	Maximum 7·8 acres per picker

Notes

* For machine working rates see Table A.1.

Source: Mainly Farm Planning Data 1966 – J. B. Hardaker[11]. Costs revised 1974.

Table A.9(i). Annual Tractor Requirements

Cash crops	Tractor hours per acre		Forage crops	Tractor hours per acre		Livestock†	Standard tractor hours per head
	Standard tractor hours	High capacity machines		Standard tractor hours	High capacity machines		
Winter wheat – combine*	6·5	4·5	Mangolds	25·0	—	Pigs – sow	2·5
Spring wheat – combine*	6·5	4·4	Bare fallow	8·0	4·0	porker	0·25
Barley – combine*	5·5	4·0	Kale – grazed	9·0	—	baconer	0·5
Oats – combine*	5·5	4·0	Kale – carted	29·0	—	heavy hog	0·5
Cereals harvested by binder add	3·5	—	Establishment of leys – undersown	1·0	—	Poultry – layers	0·05
Sugar beet – mechanical harvest	29·0	19·0	direct seed	4·0	—		
Sugar beet – complete mechanical	32·0	21·0				Sheep – per ewe	1·5
Potatoes – hand harvest	25·0	20·0				Store sheep	1·0
Potatoes – mechanical harvest	30·0	21·0	Grass – hay	6·5	5·0	Dairy cows	8·0
			Grass – silage (f.h.) 1st cut	6·5	4·5		
Peas – combined dry	11·0	8·0	2nd cut	4·5	3·0	Cattle – 2 years and over	7·0
Field beans – combined	7·0	5·0	Grazing – temporary grass	2·0	—	1–2 years	5·0
Peas – picking	8·0	6·0	permanent grass	1·5	—	calves	3·0
Peas – vining	10·5	8·5	Baling and carting straw	2·0	1·0	Yarded bullocks	4·0

Tractor capacity ‡	
Required capacity in standard tractor hours	Number of small/medium tractors
Up to 800	1
801–2000	2
2001–3200	3
3201–4400	4
4401–6800	5
5601–6800	6

* Excluding baling straw.
† The figures for livestock represent servicing with food and litter from store, clamp or field.
‡ See page 9.

N.B. This table should be used only where no local figures are available. Adapted from *The Farm as a Business*, Section 6.

Source: Based on data supplied by University Agricultural Economics Departments and N.A.A.S. surveys.

Table A.9(ii). Tractor Sizes and Ratings (Standard Tractor Hours)

Tractor size	h.p.	'Rating'. Standard Tractor Hours per Hour*
Small	15–30	$\frac{2}{3}$
Small-medium	31–45	1
Medium	46–60	$1\frac{1}{4}$
Large-medium	61–80	$1\frac{2}{3}$
Large	81–100	2
Very large	Over 100	$2\frac{1}{4}$–$2\frac{1}{2}$

* Author's estimate. Depends on utilization.

Table A.10. Guide Prices for Contract Work

The following typical contract prices are based on provision of professional agricultural contracting services on about 25 acres. Prices will naturally vary from one district to another and will be affected by the amount of work done.

	per acre £	per hectare £	per hour £
Ploughing			
Grassland or arable	7·00	17·50	
Deep ploughing (over 10 in (25 cm))	15·00	37·50	
Subsoiling	3·90	9·60	
Cultivating			
Medium	2·60	6·50	
Chisel, once	5·00	12·50	
twice	7·25	18·10	
Spring-tine	2·50	6·25	
Rotary cultivating (medium tractor)			7·70
Harrowing			
Disc, standard	2·70	6·75	
Disc, heavy	3·65	9·10	
Medium zig-zag, seedbed	1·70	4·25	
Grassland	2·05	5·10	
Rolling			
Ring (set of 3)	2·00	5·00	
Flat (4-ton)	3·50	8·75	

Table A.10—*continued*

	per acre £	per hectare £	per hour £
Drilling			
Combine, farmer to cart seed and fertilizer	4·40	11·00	
Corn only, farmer to cart seed	3·40	8·50	
Grass or small seed	2·70	6·75	
Grass seed (broadcast)	1·90	4·75	
Sugar beet and roots, spacing drill	5·00	12·50	
Direct drilling grain and fertilizer, or kale	5·95	14·85	
Direct drilling grass once	5·95	14·85	
Fertilizer distributing			
Spinner up to 3 cwt/acre (375 kg/ha) (fertilizer in field)	1·25	3·60	
Granule application	1·25	3·60	
Spraying (excluding materials)			
Low volume (up to 20 gal/acre (225 l/ha))	1·80	4·50	
High volume (over 20 gal/acre)	add 3p/ gal/acre		
Low volume aerial application			
Fixed wing aircraft	2·00	5·00	
Helicopter	2·35	5·85	
Band spraying	2·00	5·00	
Grass mowing or topping	3·50	8·75	
Tedding	2·50	6·25	
Swath turning	2·20	5·50	
Forage harvesting (8 tons/acre) (20 tons/ha)			
Forage harvester and tractor			
Single- or double-chop	6·10	15·25	
Full-chop	7·60	19·00	
Tractor and 2 trailers (1 man)	5·90	14·05	
Tractor and buckrake	3·80	9·50	
Corn harvesting			
Combining, with tanker	11·85	28·60	
Combining with carting to store	15·60	39·00	

Table A.10—*continued*

	per acre £	per hectare £	per hour £
Strawchopping, tractor-drawn	3·60	9·00	
Baling			
Pick-up baling, inclusive of twine, 12 pence per bale			
Muck loading and spreading (minimum 8 hours)			
Wheeled tractor and loader			5·65
Industrial type loader			5·75
Tractor and spreader (power drive, 4-ton)			6·05
Oil-seed rape harvest			
Windrowing	8·10	20·25	
Combining with pick-up attachment	20·00	50·00	
Hedge cutting			
Circular saw or flail type (1 man)			6·05
Sawing logs (minimum charge £10)			
Bench, tractor and 1 man			4·90
Chain saw and operator			3·50
Pea cutting, green or dry	9·20	23·00	
Lime and basic slag spreading			
Kibbled/burnt/shell lime £1·50 per ton			
Ground limestone @ 2 ton/acre £1·00 per ton			
Basic slag £3·00 per ton			
Tractor hire (minimum 8 hours)			
Medium wheel, and driver (75 h.p.)			5·15
4-wheel-drive, and driver (75 h.p.)			5·25
(120 h.p.)			5·85
Crawler, 65 h.p. with 4-in-1 bucket (plus transport)			5·95

Source: National Association of Agricultural Contractors.

Table A.11. Approximate Metric Conversions

	Imperial		Metric/SI
Length			
	Inch	1 in = 25·4 mm	millimetre
	Foot	1 ft = 0·3048 m	metre
	Yard	1 yd = 0·9144 m	metre
	Mile	1 mile = 1·609 km	kilometre
Area			
	Square inch	1 in² = 645 mm²	square millimetre
	Square foot	1 ft² = 0·09 m²	square metre
	Square yard	1 yd² = 0·84 m²	square metre
	Acre	1 acre = 0·4047 ha	hectare
Volume			
	Cubic foot	1 ft³ = 28·3 dm³	cubic decimetre
	Cubic yard	1 yd³ = 0·76 m³	cubic metre
	Gallon	1 gal = 4·55 l	litre
Mass			
	Pound	1 lb = 0·454 kg	kilogram
	Hundredweight	1 cwt = 50·8 kg	kilogram
	Ton	1 ton = 1016 kg	kilogram
		= 1·016 t	tonne
Velocity			
	Feet/minute	1 ft/min = 0·0051 m/s	metre/second
	Mile/hour	1 m.p.h. = 1·61 km/h	kilometre/hour
Force			
	Pound force	1 lbf = 4·448 N	newton
Pressure			
	Pound/square inch	1 lbf/in² = 0·06895 bar	bar
		= 0·07 kg/cm²	kilogram/square centimetre
	Inch mercury	1 inHg = 33·86 mbar	millibar
Power			
	Horsepower	1 hp = 0·746 kW	kilowatt
Rate of use			
	Ton/acre	1 ton/acre = 2510·7 kg/ha	kilogram/hectare
	Hundredweight per acre	1 cwt/acre = 125·5 kg/ha	kilogram/hectare
	Acre-inch	1 acre-inch = 254 m³/ha	cubic metre/hectare
	Gallon per acre	1 gal/acre = 11·2 l/ha	litre/hectare
Heat, insulation, ventilation			
	British thermal unit	1 Btu = 1·055 kJ	kilojoule
	Calorie	1 cal = 4·1868 J	joule
U-value			
		1 Btu/ft² h deg F = 5·678 J/m² s deg C	
K-value			
		1 Btu/in/ft² h deg F = 0·144 J/m/m² s deg C	

Index